A TOUCHSTONE BOOK
Published by Simon & Schuster, Inc.
New York   London   Toronto   Sydney   Tokyo

THE SECRET HOUSE

24 · HOURS · IN · THE STRANGE · AND UNEXPECTED · WORLD · IN WHICH · WE · SPEND · OUR NIGHTS · AND · DAYS

DAVID BODANIS

Copyright © 1986 by David Bodanis

All rights reserved
including the right of reproduction
in whole or in part in any form.

First Touchstone Edition, 1988

Published by Simon & Schuster, Inc.
Simon & Schuster Building
Rockefeller Center
1230 Avenue of the Americas
New York, NY 10020

Originally published in Great Britain by Sidgwick & Jackson Ltd., 1986

TOUCHSTONE and colophon are registered trademarks
of Simon & Schuster, Inc.

Designed by Bonni Leon

Manufactured in the United States of America

10 9 8 7 6 5 4 3
10 9 8 7 6 5 4 3 2 1 Pbk.

Library of Congress Cataloging in Publication Data

Bodanis, David.
  The secret house.
  1. Science—Miscellanea. I. Title.
Q173.B66 1986   500   86-13036

ISBN: 0-671-60032-X
ISBN: 0-671-65718-6 Pbk.

CONTENTS

INTRODUCTION & ACKNOWLEDGEMENTS

I got the idea for this book when I was living for some time in a curious house in a small French village. The ground floor was built around a 12th century Saracen wall, the next level was a few centuries older, and the whole thing worked its way up to a fourth floor terrace c. A.D. 1978. Each level had a different feel, a different psychology. What would it be like to work out this psychology for the contemporary home?

A fortuitous rereading of Husserl at this point suggested the approach I wanted to take. Why not describe the immediate environment we're surrounded by as we struggle or frolic through a typical day? The only problem was the tone. I spent a few months trying draft chapters, assembling mounds of crumpled paper, wads prodigious enough to cause excitement among the locals, before ending up with something that seemed right. It turned out to be a sort of benevolent impersonality. I am benevolent, though the facts stay impersonal. Others have gone that way before – Tati, Swift and Gimpel would be at home here – but I liked the twist of doing it in scientific terms, appropriate to the setting of the thoroughly modern man.

Once that was straight there remained only the small matter of researching and writing the full book. Small French villages have many advantages, but facilities for science research are not among them. London looked a better bet. June Hall, my agent, got the necessary funds for this stage, having found in William Armstrong an adventurous publisher. Without his encouragement this book would probably not have been done.

Once in London the libraries made research easy. The Science Reference Division of the British Library, the Imperial College Library, and the main University of London Library were a pleasure to use. For historical matters, the Science Museum Library was tops.

Perhaps an even nicer surprise were all the organizations and individuals willing to discuss minute problems of applied science or technology with an often ill-informed questioner who had only a purely theoretical background. These included: The Air Infiltration Centre; The Assay Office, Goldsmiths Hall; Dr J.P. Blakeman, Queen's University, Belfast; Dr Sally Bloomfield, Dept of Pharmacy, Chelsea College; Brick Advisory Centre; British Association for Chemical Specialties; British Institute of Cleaning Science; British Mycological Society; British Plastics Federation; British Textile Confederation; Building Research Station; Dr Burtonwood, Huddersfield Poly; Dr J.M. Clark, Porton Down Centre for Applied Microbiology and Research; Prof. Peter Clarke and his colleagues at the Building Dept, Trent Poly in Nottingham; Cosmetic Toiletry and Perfumery Association; Electricity Council; Dr George East, Leeds University; Dr Griffiths, Pest Infiltration Laboratory; Dr Alan Hedge, Environmental Psychology, University of Aston; Prof. Ewins and Dr Flower at Imperial College; Dr Peter Jackman, Air Infiltration Centre; Dr Peter Jonas, Meteorological Office, Bracknell; Dr H.G. Leventhal, Head of Acoustics, Atkins Research and Development; Dr O.M. Lidwell, Common Cold Unit, Harvard Hospital Salisbury; Dr C.A. Mackintosh, Central Public Health Laboratory; Ministry of Agriculture, Fisheries and Food; Dr John Ridgeway, Water Research Centre; Mr M.J.V. Powell, Construction Industry Research and Information Board; Rentokil Ltd; Royal Institute of British Architects; Royal Institute of Public Health and Hygiene; Royal National Rose Society; Sainsbury's Technical Laboratories; Schweppes Research and Development Division; at the Shirley Institute, Drs Wilson and Jeffries; Timber Research and Development Establishment; Society for Applied Bacteriology; Southern Gas; Dr David Whitehouse, Mullard Space Science Centre. None of them of course are responsible for the errors of fact or interpretation that remain.

Michael Marten of the Science Photo Library, together with his team of professional photographers and international contacts, generously helped to research and supply the pictures.

The organizational and line-editing was carried out in London by Libby Joy and Carey Smith, and in New York by Donald Hutter. If all editors were cloned in their image, writers would have nothing to complain about.

This introduction is being completed at five o'clock in a London morning. After delivering it to the publishers the author will return to his curious multi-level French house. How it will look now remains to be seen.

PART ONE

DAYTIME

Picture taken with infra-red film revealing heat patterns in an average house at night. White is the hottest; blue and black coolest. Note the massive heat loss from the roof, due to billowing convection clouds of warmed air inside

ONE
MORNING

From the alarm clock a spherical shock wave traveling at Mach 1 starts growing outward, spreading and spreading till it hits the wall. Some of the energy it carries causes the curtains over the window to heat up from the friction of the onslaught; much of the rest rebounds back, enters the ears of two sleepers, and finally rouses them awake.

There's a rolling of eyes and a stirring of head, then a female hand gropes out from under the security of the comforter, fumbles on the bedside table, finds the alarm clock, and clacks down the button on top to turn it off.

The buzzing from the alarm clock stops, but the even higher-frequency shriek from the quartz crystal inside takes over, spreading in a growing sphere out from the clock as the sound wave did, striking the walls and heating the curtains too. But this second room-filling shock wave is inaudible. The waker, desperate to fill the rigors of the morning with some soothing music, fumbles out from the covers again towards the radio. The spherical pulses from the quartz in the alarm clock crash unheeded over her arm, but she is not to be deterred: the radio is found, switched on, listened to for a brief instant, then the tuning knob is furiously grasped.

Some simpleton had left it on the news station last night. Now it must be moved from that drivel and switched to the haven of the classical music station.

The tuning knob quickly rolls, speeding across the megahertz to the new location. There's a crackling as it moves between stations; a slight hiss and buzzing too which the waker single-mindedly ignores. Certain of the hissings are the cries of distant exploding galaxies, consumed in their death throes and sending out massively powerful particle radiation across space and time in the process of obliteration. Other static comes from lightning strikes on distant continents, which send electro-magnetic pulses through the upper atmosphere that travel across deserts and seas into the bedside radio; all are received, then passed over and ignored in the hunt for the right station.

The radio disturbs the other sleeper, and after some fruitless tugging of the comforter, a division of resources is ordained. The first waker lies back to savor the music, while the second one, stifling deep inner protest, unsure of his domain and wondering where the civilized conversation of his usual radio station has gone, gets ready to emerge from the bed.

Whack thump bam! The man's foot extends out of bed and lands on the floor. The floorboards jam down and their vibrations travel sideways like pond waves to the wall. The whole house compresses in the new loading — bricks where the floor fits into the wall shrinking smaller by $1/100,000$ inch from the weight.

Any impact that doesn't get lost in the walls stays quivering in the floor. The chest of drawers starts lifting up and down, as does the bed, the chair, the table with its plant on top, the stack of magazines and Sunday papers in the corner, and even the old coffee cup left down on the floor. All lift up and bounce down, rebound up and crash down again, as the floor reverberates to get rid of its buzzing energy. In a particularly energetic leap out of bed this bouncing of furniture can be seen (lampshades especially are prone to being knocked over in such moments), but even with a softer landing the furniture shaking takes place.

Then the second foot touches down, the waker stands up, and he steps to the double-glazed window to see what is happening outside.

It is, as usual, raining. Not water raindrops — that's only on stormy days. This is an electric rain uncovered first thing in the morning, a rain of charged air particles that started as simple decay products from radioactive gas nearby. (House walls spray out radioactive gas — a lot if it's brick or concrete, less if it's wood or metal-clad — and front walks and street surfaces do the same.) The particles have been hovering in the lower atmosphere's invisible electric field ever since. This electric rain spatters the lawn, the front walk, the roof, and now it sprays in through the open window. It's a gentle rain — perhaps 200 volts per yard, but at a tiny amperage — and it dribbles away in speckles of leaking charge on the bedroom walls wherever it hits.

The window in its aluminum frame is slid closed; the invisible shower has not proven of captivating interest. As the window closes, long slivers of its aluminum frame come tearing off. On a steel window such tearing friction would provide a nice niche for future rust to sprout. But here the aluminum window quietly goes about repairing the scratch itself, as it always has. Before the culprit has even turned away a new layer of aluminum oxide starts growing out sideways from the intact portion. It spreads across the microscopic gouge, covers it and seals it, and only stops when it has formed a perfectly fitting replacement for the bit that was lost.

When aluminum was first discovered in the last century such self-repair caused it to be treated as only a precious and mysterious new metal could expect to be. The Emperor of France scrapped his pure silver dining set and had it replaced by one composed entirely of this new and vastly expensive material. The American Congress considered a proposal to buy some of this glittering substance and put it in a slab on top of the Washington monument as a sign of respect to the nation's founder; the project was subject to lengthy discussion because of the outrageous cost. Only when British chemists discovered aluminum alloy in a downed German Zeppelin in the First World War did its use as an ordinary metal begin to spread, so that now its presence in low-cost sliding windows is treated as perfectly normal.

The waker is now on his way to the bathroom. As he steps, the floor continues to shake, and the dust continues to dance from the invisibly rebounding furniture. But there's also something else which moves under his feet, some *things* rather, roused out of their sleep as the waker strides over them.

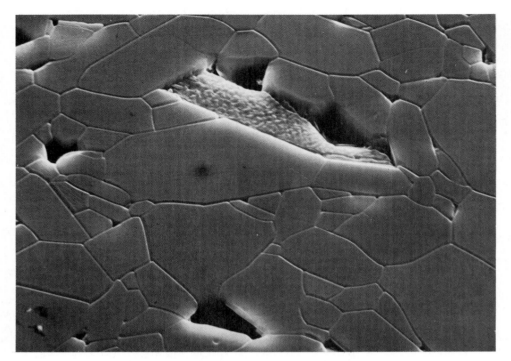

Aluminum, magnified 4000 times. Even this sample, purer than that in most sliding windows, reveals gaps, cracks, holes and impurities

These are the mites, thousands and thousands of tiny mites: male mites and female mites and baby mites and even, crunched to the side away from the main conglomerations, the mummified corpses of long-dead old great-grandparent mites. Brethren of theirs stir in the bed too, where they have spent the night snuggling warm and cosy under our sleepers, and which now, the great burden above them stirring, are beginning to stir for the day too.

It sounds unpleasant, but is quite normal. You don't have to leave the same sheets on for weeks, let the dog crawl everywhere, and generally do all those other awful, unhealthy things we expect of people whose rooms are infested with bugs to get them. Even if the room is well aired and the floor clean – the dog never, ever let up to play – the mites will still be there. Epidemiological studies show that nearly 100 per cent of our houses are host to these creatures. It's the same in Germany, Sweden, and apparently every other advanced country too. The consolation is that these are not great visible mites that produce itching, let alone the all too visible and loathsome bed-bug, but rather a special, ultra-tiny (so small they

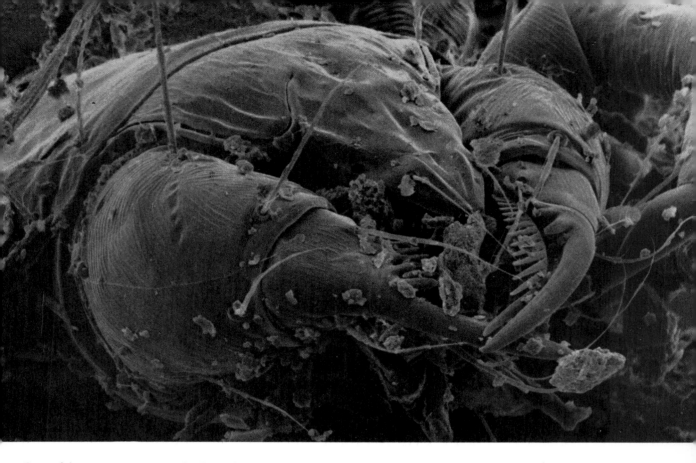

One of the many usually sub-visible dust mites in your home, here magnified 1000 times. Note the ecologically useful body details, such as serrated front claws (for collecting flakes of human skin) and the protective body armor. The dust mite is a gentle, passive creature

were only first discovered in 1965) breed that lives in human carpets and beds and nowhere else.

Mites have been called sacs with legs on, and that's a fair description. There's a mostly naked body, a few loose armor plates on it, holes for breathing, eating, elimination, and copulation, and stubby little hairs sticking out all over to help feel what's going on. Each one has eight legs, because at one time they were in the same evolutionary line as the spider, but that was over 300 million years ago, and since then matters have changed. The spiders went on to be great multi-eyed hunting carnivores; the mites went a different way, and many have ended up as peaceful grazers, munching whatever is left over from the larger creatures they shelter near.

In the house these leftover nibbles are skin: tiny rafts of human skin flakes. There's plenty of it around. It's rubbed off when you move in bed, and it's brushed off when you dress. It falls off the body at a stupendous rate whenever you walk — tens of thousands of skin flakes per minute — and it tears off at only slower rates when you stand perfectly still. For us the skin flakes are insignificant, noticeable only when they build up as dust, but for the waiting mites they are manna.

Hidden down at the base of the carpets these mites only have to wait, mouth up, for this perpetual haze of skin flakes to rain down on them – the ultimate in parachuted food rations. For the mites that live in the bed (an estimated 42,000 per ounce of mattress dust; 2,000,000 total in the average double bed) the floating skin rafts are even more accessible. They slip through the weave of any pyjamas that might be worn, through the spaces between individual threads in the bottom sheet, and so tumble down onto the mites contentedly waiting on the bottom. The warmth in the bed is nice, for the mites' original evolution was in the tropics, but carpets are all right too because the mites just slow down all their actions to make do in the greater cool in carpet fabrics.

What the mites do in their protected habitats is what most animals spend their earthly existence doing. They eat, they defecate, and in propitious moments they copulate. Twenty faecal pellets emerge from each mite a day, squeezed out of special anal valves. A vast heap containing nearly as many pellets as there are stones in the Great Pyramid would fit easily on the period at the end of this sentence (and a few are probably already there, come to that). The faecal pellets are so small that they float, an ascending offering-up to the gods who kindly let loose the sustaining skin flakes perhaps, and they soar and travel throughout the house.

Some of the mummified ex-mites are hollow and light enough to float up too – another Egyptian-style funerary offering to join the pellets. Where the mites make things more complicated than the ancient Egyptians is that not all floating mite-shaped husks in the home are mummies. Some are just the discarded shells of growing mites, for like many insects these carpet and bed dwellers shed their skin on a regular schedule: having it go dry, crack open, and a new naked mite step out.

A half day or so after being reborn this way the fresh mites are ready to mate. It is a delicate process. In certain cases the male produces a sealed packet of sperm, leaves it on a convenient surface, and then departs. The female, chastely having no part in the proceedings up to that point, then discreetly sits on the packet, or, in the case of those females with the genital openings on their top surfaces, plops backwards on it.

It's not quite what we're used to, but it works. Mite families with thousands of members have been found living 16,000 feet up on Mt Everest; others have been found in the Antarctic, deep under the Pacific Ocean and even, in the case of the one New Guinea species, living out their whole lives – successful mating procedures included – within the fungal growths that are carried on the backs of large weevils living in moss forests. The environment of suburban bed and floor present few rigors after that.

Into the bathroom goes our male resident, and after the most pressing need is satisfied it's time to brush the teeth. The tube of toothpaste is squeezed, its pinched metal seams are splayed, pressure waves are generated inside, and the paste begins to flow. But what's in this toothpaste, so carefully being extruded out?

Water mostly, 30 to 45 per cent in most brands: ordinary, everyday simple tap water. It's there because people like to have a big gob of toothpaste to spread on the brush, and water is the cheapest stuff there is when it comes to making big gobs. Dripping a bit from the tap onto your brush would cost virtually nothing; whipped in with the rest of the toothpaste the manufacturers can sell it at a neat and accountant-pleasing $2 per pound equivalent. Toothpaste manufacture is a very lucrative occupation.

Second to water in quantity is chalk: exactly the same material that schoolteachers use to write on blackboards. It is collected from the crushed remains of long-dead ocean creatures. In the Cretaceous seas chalk particles served as part of the wickedly sharp outer skeleton that these creatures had to wrap around themselves to keep from getting chomped by all the slightly larger other ocean creatures they met. Their massed graves are our present chalk deposits.

The individual chalk particles – the size of the smallest mud particles in your garden – have kept their toughness over the aeons, and now on the toothbrush they'll need it. The enamel outer coating of the tooth they'll have to face is the hardest substance in the body – tougher than skull, or bone, or nail. Only the chalk particles in toothpaste can successfully grind into the teeth during brushing, ripping off the surface layers like an abrading wheel grinding down a boulder in a quarry.

The craters, slashes, and channels that the chalk tears into the teeth will also remove a certain amount of built-up yellow in the carnage, and it is for that polishing function that it's there. A certain amount of unduly enlarged extra-abrasive chalk fragments tear such cavernous pits into the teeth that future decay bacteria will be able to bunker down there and thrive; the quality control people find it almost impossible to screen out these errant super-chalk pieces, and government regulations allow them to stay in.

In case even the gouging doesn't get all the yellow off, another substance is worked into the toothpaste cream. This is titanium dioxide. It comes in tiny spheres, and it's the stuff bobbing around in white wall paint to make it come out white. Splashed around onto your teeth during the brushing it coats much of the yellow that remains. Being water soluble it leaks off in the next few hours and is swallowed, but at least for the quick glance up in the mirror after finishing it will make the user think his teeth are truly white. Some manufacturers add optical whitening dyes — the stuff more commonly found in washing machine bleach — to make extra sure that that glance in the mirror shows reassuring white.

These ingredients alone would not make a very attractive concoction. They would stick in the tube like a sloppy white plastic lump, hard to squeeze out as well as revolting to the touch. Few consumers would savor rubbing in a mixture of water, ground-up blackboard chalk and the whitener from latex paint first thing in the morning. To get around that finicky distaste the manufacturers have mixed in a host of other goodies.

To keep the glop from drying out, a mixture including glycerine glycol — related to the most common car anti-freeze ingredient — is whipped in with the chalk and water, and to give *that* concoction a bit of substance (all we really have so far is wet colored chalk) a large helping is added of gummy molecules from the seaweed *Chondrus Crispus*. This seaweed ooze spreads in among the chalk, paint and anti-freeze, then stretches itself in all directions to hold the whole mass together. A bit of paraffin oil (the fuel that flickers in camping lamps) is pumped in with it to help the moss ooze keep the whole substance smooth.

With the glycol, ooze and paraffin we're almost there. Only two major chemicals are left to make the refreshing, cleansing substance we know as toothpaste. The ingredients so far are fine for cleaning, but they wouldn't make much of the satisfying foam we have come to expect in the morning brushing.

To remedy that every toothpaste on the market has a big dollop of detergent added too. You've seen the suds detergent will make in a washing machine. The same substance added here will duplicate that inside the mouth. It's not particularly necessary, but it sells.

The only problem is that by itself this ingredient tastes, well, too like detergent. It's horribly bitter and harsh. The chalk put in toothpaste is pretty foul-tasting too for that matter. It's to get around that gustatory discomfort that the manufacturers put in the ingredient they tout perhaps the most of all. This is the flavoring, and it has to be strong. Double rectified peppermint oil is used — a flavorer so powerful that chemists know better than to sniff it in the raw state in the laboratory. Menthol crystals and saccharin or other sugar simulators are added to complete the camouflage operation.

Is that it? Chalk, water, paint, seaweed, anti-freeze, paraffin oil, detergent and peppermint? Not quite. A mix like that would be irresistible to the hundreds of thousands of individual bacteria lying on the surface of even an immaculately cleaned bathroom sink. They would get in, float in the water bubbles, ingest the ooze and paraffin, maybe even spray out enzymes to break down the chalk. The result would be an uninviting mess. The way manufacturers avoid that final obstacle is by putting something in to kill the bacteria. Something good and strong is needed, something that will zap any accidentally intrudant bacteria into oblivion. And that something is formaldehyde — the disinfectant used in anatomy labs.

So it's chalk, water, paint, seaweed, anti-freeze, paraffin oil, detergent, peppermint, formaldehyde and fluoride (which can go some way towards preserving children's teeth) — that's the usual mixture raised to the mouth on the toothbrush for a fresh morning's clean. If it sounds too unfortunate, take heart. Studies show that thorough brushing with just plain water will often do as good a job.

And anyway, a much more pleasant mixture of substances for the mouth awaits in that other chamber of chemical concoctions a little way down the hall.

Come in the kitchen, toss the newspaper on the table, and what's there to see? A great mob of tiny creatures fleeing for their lives. These are the pseudomonads, and they're one of the most widespread bacterial beings in the home. There are many thousands hanging on by strands to your face as you walk in, a good number on your arms and bathrobe, and even more on any recently moist house surface.

Here on the kitchen table they're flailing and twisting and desperately trying to make it to the far side, but in this endeavor they suffer one near-insurmountable problem. They're too small. Not the size of a mite, not the size of half a mite, but smaller still, itty bitty, so tiny that a pile of a hundred thousand would not be visible, and so short that the far end of the table, the one they're lunging for, is an awful 400,000 body lengths away — what for us would be 400 miles over a flat road.

And the creatures can't walk. They can't walk, trot, jog, skip, or sprint; they can't hop; they can't even crawl. With no arms or legs, all those modes of locomotion are out of consideration. They're shaped like simple stubby tubes, what we might look like if stuffed into a rubber wetsuit that had no outlets for our limbs or head. Yet even shaped like that, these pseudomonads still have to flee across the table from this awful thing that is coming after them.

So they swim. No arms, so the breaststroke is out; no legs, so a flutter kick wouldn't help; but something even better than that is provided for their use. A propeller. Each one of the great mass of creatures congregating on the kitchen table, aged elders and just arrived young-sters, has a massive, powerful tail sticking out of its rear end. It doesn't have any muscles in it, but it does have an angled pipe attachment at the base leading inside, and where that fits there is a chemical motor in position to drive it.

As it's their only way forward, it has to be able to start every time. It does. Each creature just aims, and the motor starts up, the propeller begins to turn, and it's off. Round and round the propeller goes, and forward and straight the creature is propelled. To us its speed would not be particularly impressive — 0.0001 miles per hour — but on the scale of these little beings, that's fast. Scaled down to their length it equals seven body lengths a second, which is better than even Carl Lewis at his best. Still, even an Olympic sprinter would take a long while to make 400 miles, and these table-top creatures, despite having their propeller going for them, have a long haul ahead.

And they conduct it in a strange way. Each creature starts off okay, moving forward steady and on course, propeller thumping, but as soon as it's made any amount of progress it suddenly goes berserk. The turning propeller stops turning, goes still, and then starts up in the opposite direction.

That hauls the creature back in reverse, twisting and yawing and veering it off its course, as an outboard motor switched into reverse on a fishing boat would do. But then, just as suddenly, the aberration ends, the tumbling and twisting stop, the chemical motor turns the propeller back on the right way. The creature advances forward again, with no sign at all that this little hiccup ever happened.

The whole reversing ruckus only came about because the creature was trying to get its bearings. It's a pretty peculiar way of getting your bearings, but in lieu of compasses or sun angles it's the best they can be expected to do. Also, it works. If the creature propels itself into a danger zone there will be lots of these moments of backward twisting, and that will help to get it turned around heading to safety. If it's already in a safer region there are fewer slippages – just enough to check that everything's still okay – and the creature stays headed the right way.

So the mob progresses along the table, with backslidings and confusions and collisions and an unholy mess to be sure, but still moving away from that *awful* thing which it would be very unwise to stay near.

This thing is just the morning newspaper, tossed so offhandedly onto the kitchen table a moment before. As the paper impacted great gobs of material in it fell out. Not the newsprint, that's rigid enough, but such extras as the hemp, linen, wool, asbestos, glass fibres, glues and other substances the modern newspaper is filled out with. And in that deluge there was also an extraordinary gush of loose ink (usually microscopic, though sometimes slightly inky fingers after the crossword stage has been reached give an intimation of what's been happening).

Newspaper ink is not attached to the fibres of the paper, but instead is held on by rough cracks in the interstices between them. The slamming impact of a newspaper on the table is enough to rub some of that loose. It splashes out in horizontal jets, and any pseudomonad creatures caught in

the advancing dark fluid would not have long to live. Aside from the oils and detergents in the ink there is also a high concentration of antimicrobial poisons worked in. It's good for keeping ink stores fresh when they're waiting at the printers, but here, splattering loose on the table, it's what causes the frantic riot of fleeing pseudomonads we started with.

So far so good. But how did the pseudomonad creatures get to be on the table in the first place? They came from the dish towel in the kitchen, and the horrors contained within that simply-striped object are not for the weak-hearted to behold.

Newspaper seen close up, as fragments of wood chips, wool, asbestos, glass fragments, linen, and other constituents. The subject of the half-tone photo (right) is Mrs Thatcher

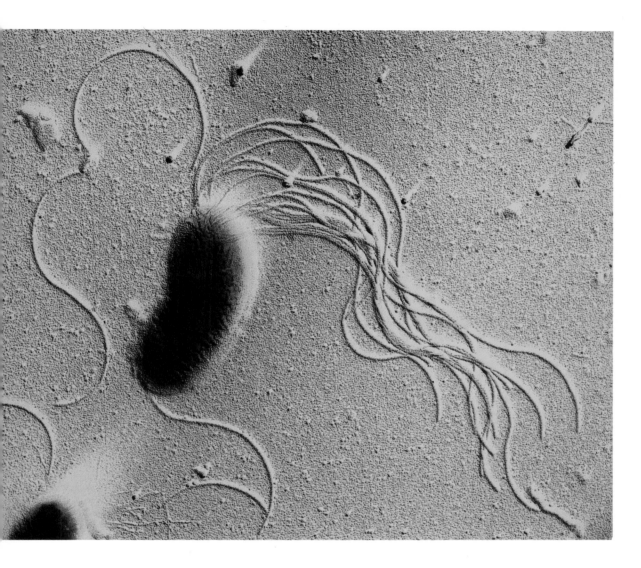

Pseudomonad bacterium, of the type that swims on your kitchen table and in damp sponges. The extending strands spin like a twirling lasso to propel the creature along

What is a dish towel? A simple weave of cotton fabric. That means, as we shall see in discussing clothing, that it has long protoplasm-filled tunnels winding within it, with minerals, proteins and soluble fats sprinkled on, and a nice sugary-based cellulose gridwork to hold the whole thing together. Bacteria would munch happily on that by itself, but since it's a dish towel there are two other additions mixed in.

One is food: the odd fragment of bread crumb, a bit of grease left on one of the washed-up plates — that sort of thing. Tucked safely away inside a dish towel, each such food residue will suffice for thousands of specialized microbes. But the second special characteristic of a dish towel is worse. That is the moisture. Bacteria thrive in moisture.

It should be no surprise then that public health microbiologists consider the dish towel (and its partner in crime, the improperly squeezed-out sponge) to be one of the leading spreaders of bacterial populations in the home. Wiped casually over the table after cleaning up the night before it deposited great numbers of pseudomonads there. Moisture transferred with them, and the nutrients dissolved in that moisture provided all that the creatures needed to multiply for the morning after. The problem could be avoided by wiping down only with a fresh cloth, or making sure every surface is bone dry before finishing, but who's inclined to see to that every evening?

All this buggy propagation can give the wrong impression about what's been going on in the kitchen overnight. Certainly there are all those creatures on the table, and many more where they came from in the sink, fridge, and on the shelves. But that does not mean they have been assaulting the stored food all night, swimming up to it, squeezing it, munching it, and generally making a mess of it. They've tried, but they've often failed to make even a mark. For the food in your kitchen fights back.

Consider the humble egg within the refrigerator. Overnight it was steadily breathing, wheezing out and pulling in all the gases in the refrigerator atmosphere. It did its breathing through little holes on its surface, and while those holes might be just the thing to bring in oxygen for any embryo that was to form within the egg, they also have the bad habit of letting in any bacteria – often cousins of the same pseudomonads we met on the table – that happen to have ended up on the surface of the egg. The holes are shaped like hollow golf tees, and are broad enough for a dozen or more mooching bacteria to slide down each one at a time.

So much for the weak point. Now for the defense. At the bottom of those holes the bacteria do not get a simple swim across to the yolk where the nutrients they could use are located. There's something in the way. Right at the bottom of each hole, a hard rubbery membrane is stretched across. (In hard boiled eggs it's noticeable as the translucent film around the egg.) Getting through is no easy job. The bacteria have to twist and probe, secrete dissolving enzymes and then push some more before they can pierce a hole through this stage one barrier. And considering the fate that lies in store for them on the other side, it might be better if they didn't make it.

To us the white of an egg is a simple gooey watery thing that's interesting because it turns hard on being cooked, and makes a nice contrast to the yellow yolk. Otherwise it doesn't seem much good for anything. But on the microscopic level it is something more. The white surrounds the yolk like a chemically loaded and booby-trapped sea, quite capable of destroying fleets of bacteria that enter into it on their voyage across to the yolk.

The invading bacteria are allowed to slip into this sea without problem once they've made it through the first barrier. They're even given a few minutes of unbothered propeller-throbbing swim. Only then, well away from the safety of the shell wall, does the attack begin. Sizzling lysozyme enzymes sear into the bacteria's cell walls, rending them open and leaving great numbers dead. More lysozyme streams on the bacteria, and then even more before the sizzling lets up. (The same lysozyme defender is found in human tears, there regularly splitting open errant bacteria too, and so leaving the surface of the eye one of the few regions of the body relatively bacteria free.)

After the lysozyme attack, the way might look clear for the bacteria survivors, but in fact things have only become worse. All around them the nutrients that they need to sustain them on the swim are slowly being wrapped up by living processes in this strange sea of white. The bacteria need iron to keep their propellers turning; the egg white creates slithering proteins that latch onto the nearest iron before the bacteria do, wrap around it, and chemically lock it away. Zinc and copper nutrients get wrapped up by this insidious protector as well.

Possibly the bacteria could make it anyway, for even without the main iron meals they need, it's just possible that their propellers might have enough thump left to push them across. If the bacteria could just swim over to the nearest vitamins (riboflavin and biotin are especially abundant) just a few body lengths away, they might still be able to cross in time before they founder from lack of iron.

They don't make it. The living egg white creates from its depths other substances, far smaller and more manoeuvrable than the bacteria, and these reach the bobbing vitamins first, wrap around and almost destroy them, rendering them useless to the bacteria when they do finally arrive.

That was the last chance the bacteria had. Now they're too far from the shell they fell down to make it back, and in the other direction the yolk that they were aiming for all along, that would give them nice juicy proteins and fats if they could only get to it for some hunger-appeasing sustenance, that yolk is too far away in the forward direction.

As a result the bacteria die, one after the other, bobbing uselessly, their propellers still, starving from lack of food out in this unsuspected guardian sea which came alive and thwarted them once they entered it. The white, once again, has done its duty.

The tin cans in the kitchen do an equally stalwart job of preservation. Most of each can is simple pressed steel. The tin that gives it the name is a coating on the inside only a fraction of a millimeter thick. But that slender layer is enough. The tin splutters out extra electrons which form a barrier against corrosive acid in the food weakening the can, and without that corrosion there will be no microcracks or thin points for microbes to get in. Tins of mutton brought to the Antarctic by Shackleton in 1908 were opened fifty years later and found to be perfectly edible.

The mechanism is so simple, the cans are so widespread (over one hundred billion annual world production), that it's hard to imagine what a difference their introduction has made. Until the era of tin cans armies were limited in size by the number of chickens, steers, cows and other food animals they could lug along behind them, or hope to find on the way. In 1795 the French revolutionary government announced a 12,000 franc prize for a way of storing food to get around this limitation, and in 1809 a Parisian candy-maker, François Appert, pocketed the sum with an early version of the sealed can. Napoleon used sealed cans first, to supply his Grand Army on the invasion of Russia. The invasion ended in disaster but the cans of food held up fine, and the technique has only progressed in popularity since.

Finally, but in a different way defending itself in the overnight kitchen, is the gentle apple on the shelf. It has been injected with an ageing chemical by a certain outside force, and unless it can do something against that it will no longer be the firm inviting fruit it was last night. Instead its tissues will break down, its water will leak out (a good apple is

This weevil, seen here magnified 63 times, has just burst out of a rice grain stored in a kitchen, helped in its exit by an admirably long snout, with sharp cutting blades at its tip

Right: Forest of live fungi beginning to sprout on a lemon

84 per cent water), it will shrivel and soften and rot. So the apple gets to work.

It has curious stabilizing enzymes inside that get called out to fight against the tissue rotting. Like the pectin put into home-canned jellies to make them solid, these enzymes shore up the thin internal walls of the apple, not reverting the ageing, not totally stopping it and bringing back pure youth, but rather acting as the ideal rejuvenation drink at a Swiss clinic is supposed to do, slowing down time's ravages and limiting its display.

On its outside the apple can do even more. A thick and watertight wax oozes out onto its surface spreading from tiny pores within, as unnerving to see in close-up as would be the spectacle of an explorer oozing a thick rubber wetsuit out from the pores of his own skin when conditions get too rough to go bareskin. For the apple, that coating also limits the ageing damage, keeping it from going dry, and not letting its precious and limited internal water supplies out.

But what malignant entity has forced the apple into this precarious state, has given it the chemical that switched on the unwanted ageing? The answer can be found by looking at the rest of the tray where the apple rests. The other apples lying next to it are the culprits. From their surface have come trickling out clouds of invisible ethylene gas – a potent ageing stimulator. (In large quantities it has a sweetish smell, which many hospital patients will remember, for it is also used as an anaesthetic.) When it floats to any other apple it starts the changes that lead to ageing.

It's an inevitable process, coded in their DNA, that the apples can do nothing about. Apples in the wild all trickle out this ethylene gas, and they do so as a way of ensuring that when one ripens all the others around it will ripen at the same rate. The apple we started with just happened to be resting at the center of the group of apples on the tray, and so got more of this ethylene dose than the others. Not that much more though: the waxing skin and internal shoring up is going on in the others to hold off their over-hasty decay too. Tomatoes and avocados get it even worse, forced to respire at a terrific rate as the ethylene gas produced in the kitchen drifts closer to them at night.

Ethylene is so potent a ripener of fruit that it's used by many food companies to fine-tune their product. Almost all bananas are picked when they're still green and hard, and shipped that way to the countries where they'll be eaten. Then they're placed in giant airtight metal chambers, with high pressure cannisters of ethylene gas fixed on the outside. When demand for bananas is up those tanks are switched on and the ethylene gas is fed into the chambers through rubber hoses. It ripens the bananas fast, turning them yellow in just a few hours; so fast in fact that if the gas flow is not just right they'll go too far and start decaying, with sugary islands of ageing tissue floating up into their skin – visible as brown spots.

It is in this remarkably self-preserved sanctum known as the kitchen that the actual preparation of breakfast will take place. There's the egg to fry and the bread to toast, but before any of those tasks it is time for the most crucial of all breakfast substances: the coffee or tea. Their world consumption is the highest of any drink – after water.

The caffinated beverage won't detain us here; it's what's going on in the kettle that's of special interest in the morning. For there a time machine is being created, one going back not just paltry centuries but all the way to the dawn of the Cretaceous period, 135 million years ago. When the dinosaurs ruled.

Inside the heating kettle conditions are much as they were in the earth's primitive oceans. It is hot, there's a lot of steam, and a furious bubbling storm is working up. We of course wouldn't be fooled by the similarity – kettles are kettles and Cretaceous seas are Cretaceous seas – but for life forms or other remnants of that ancient period that could well be different. If such objects ever did end up in the kettle they just might feel that somehow time had been reversed: that they were back in their sultry distant seas. And then they might well start to come back to shape.

In the warm lagoons and seas of the Cretaceous, where the 50-foot Brontosaurus munched seaweed and the 80-ton Diplodocus waded, there also lived great numbers of tiny floating marine creatures. Most of the time they bobbed quite peacefully, but when they died their fragments settled to the sea bottom, there to be trodden on by the undainty dinosaurs if the water was shallow enough, covered over with lime mud, and gradually lost to sight. Over the aeons they remained down there, slipping apart into individual molecules. The cool sea water that came later could do nothing to bring them together.

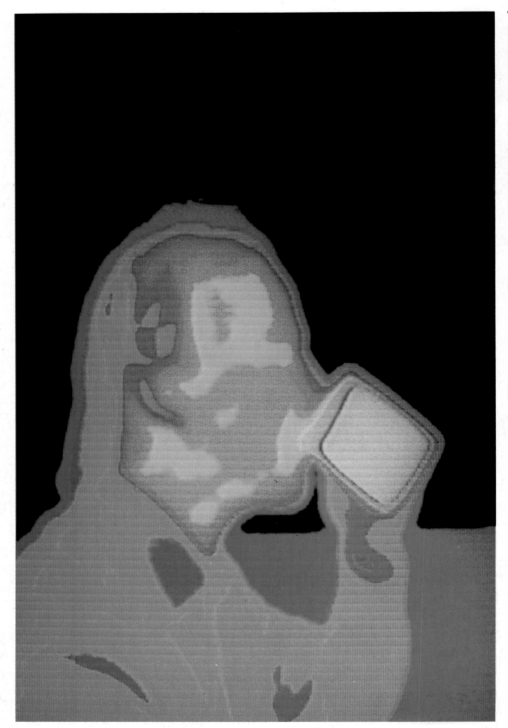

What your breakfast partner would look like if you had infra-red sensitive eyes. The coffee cup shows white from the heat, and yellow and red on the face show heat spreading there. Green and blue show clothes and hair remaining cool

Opposite: Heat picture of a kettle on the boil. The outer wall is cooler than the lid except for a weak point (yellow circle) where metal is thin. The insulated handle remains blue-violet

In time they ended up in the geological layers our present water supplies pass through. Regular amounts leach out and are carried to the house, where, just as regularly, the water is used and sent out again without these ancient chemical tombs being disturbed.

Except, that is, in the heating kettle, chugging away merrily in preparation for the needed drink. Here by contrast there is some reminder of the long-gone Cretaceous lagoons; there is the warmth and currents finally, after all this time, to make the shattered body fragments emerge out of suspension and begin to stick together again. They merge and grow and tumble together in the boiling kettle. It's not quite life: just strangely misshapen skeleton fragments suddenly reappearing after a 130-million-year hiatus. Some end up pouring into the waiting cup; some will stick to the kettle wall, building up what to our unaided eyes is an uneven furry growth, but which for these creatures is the last marker of their brief, misguided, re-emergence into the world.

Other remnants stir in the boiling kettle as well. Hardened cysts of bacterial forms that have barely changed in aeons respond to the call of the heat in there too. Their outer casings drip thin, and their inner protoplasm stirs back into life. Proteins start to be produced, bubbles move through their surface as they breathe, and, quite often, fine outer hairs start moving them through the churning currents to explore their element. It's a brief reconnoitering. With increasing heat the hairs burn off and the outer wall starts to melt. Soon the entire creature is dissolved, split apart and mixed in with the rest of the water that we will drink.

A final item stirs along with the bacterial forms and marine fragments, and it's perhaps the strangest remnant of them all. It is oxygen, but not the oxygen we're used to, not some stuff that happened to get mixed in at the distribution plant, but ancient oxygen, curious long-hidden stuff, poured out by the great fern forests back further in time, and last seen on this planet when it was breathed in by the first amphibians. It ended up chemically locked in the lime deposits on the sea bottoms along with the marine creatures, and in the same way that they responded to the heat inside the kettle, the oxygen is brought out too.

It swirls, it bubbles, it rises to the boiling froth on top and sprays out in the steam; this substance, a direct link with pre-history, floats out over the kitchen, skimming over our unnoticing breakfasters' heads, and swooping down unobserved onto their newspapers. It rises up again and some undoubtedly bears straight for their nose, where with a sigh, a sniff, or any other inhalation of breath, the link from pre-history to now is made complete.

Poured from the kettle, strained into the cup, the only thing to add to ensure the needed sustenance of tea or coffee is a simple dash of milk. Though as with almost everything in the kitchen, simple is not quite the right description. To our eyes milk is an unimpressive white fluid, but seen from the inside it is very different. Seen perhaps as a greatly miniaturized diver would see it, an enchanted living seascape is there to behold. As we can perhaps try to imagine. . . .

The first thing noticed on this scale is that the milk is not white. It's not even milky. All that registers at first is a volume of crystal-clear water (milk is 88 per cent water), extending up, down, and to all sides. It's dazzling. Then more details become clear. Glittering in the water are clusters of tiny, iridescent pebbles. These are the minerals in milk — calcium, magnesium, and others — and they're so solid almost nothing could break them. Glistening as even finer pebbles — like a school of the tiniest tropical fish perhaps — are the vitamins in milk, and these are not nearly so well built. Any sunlight penetrating down into these depths will crumble them into pieces like a laser beam, which is why you should not expose milk to direct light for long.

Quite a bit bigger, bobbing in the milk like some shipwrecked and sunken cabinet, is a hollow box-like container of casein. *It* is a pure and miraculous white. That's because it's of just the right proportions to reflect white light, which on this scale puts it about half as high as the diver is tall. Through its translucent cover several hundred even smaller cabinets can be seen nesting within.

These are the milk proteins themselves. They were packed in there at the moment of being produced in the cow. Most are just filtered nutrients from the cow's blood, but a few of these inner cabinets are the compacted corpses of bacteria and protozoa which lived in the cow's first stomach (it has four) and helped it ferment grass.

The packed container of casein is not so heavy as to sink, and it's not so light as to pop up to the surface. It hovers at a constant depth, but it does not get to hover alone for long. From the distance another stranded white cabinet bobs along. It looks like the two might touch, but then, when they're almost at arm's length, each box begins to crackle with electrostatic sparks. The closer the two cabinets drift, the more sparks there are. Soon the gap is as fiery as an arc-welder's torch spray. The incoming cabinet can't get any closer with all this commotion, and so it backs off the way it came. The sparks stop spitting, and the few left quickly quench out. This happens in all the milk you've had.

Then, off to one side, the first speck of fat bulges up from the deep. It's a hideous yellow boulder, the size of a truck on this scale, but it's so full of grease, and so much lighter than the water around it, that foaming bubbles stream off from the disturbance it makes as it rises.

In raw milk the surging fat boulders would be even larger than this — more like the size of small office blocks on our scale. Their massed arrival on the surface is what gives that milk a top coating of cream. Homogenized milk as in this pint however has only smaller fat globs – the big ones are jammed through a microscopic sieve till they splay apart.

The danger of homogenized milk is that it can make users think they're having some sort of non-fat healthier food. This of course is not the case; forcing the fat into smaller blobs does not make it go away. Even 'low-fat' milk has hundreds of these unpleasant yellow boulders in every drop poured into the coffee or tea. 'Skimmed' milk is what needs to be purchased if these fat globs are to be banned from the table.

So much for the inanimate objects in milk. There's also something else nosing about: the bacteria. If they've made it to your face, robe, toothpaste tube, kitchen table and refrigerator, they're certainly going to be able to find a niche in the milk. They're there even though the milk has been heated in pasteurization, but at this stage only in tiny numbers. There are but a few million in a pint of Grade A milk, and on the scale we've been observing that would put each floating bacteria school half a mile or more from the next.

At first the milk keeps them under control in almost the same way the egg white did. It comes alive. Any iron or vitamins on the loose get uncannily wrapped so the bacteria can't get at them; as an added defense the milk on your table actually generates pure hydrogen peroxide (in small quantities) to burn the ones that don't have the grace to starve.

For several days it will be a stalemate: the few million bacteria in a pint of Grade A squeezing tight and letting out new bacteria; the milk's defenses swooping in and starving or scalding them away. Only after a week or so will the bacteria's constant reproductive efforts win out. First little ping pong ball-shaped ones start to multiply, spreading and growing and plopping young ball bacteria out all over the water. Then jelly-coated cigar-shaped bacteria start propagating in your milk, latching on to each other in long lines, and spreading their little progeny throughout the water in turn. The water turns pretty sour with all these new bacteria dribbling their waste products into it. That sharp taste is one of the first sure signs that the milk is finally being swamped.

Then the bacteria stop the casein boxes from generating electrostatic sparks to push away from each other. Any two floating boxes that touch will stick. Soon there are submerged islands built up entirely of these boxes, and at that point they become light enough to float. Up they go, bumping and adhering even more as they rise, forming super-islands, casein continents, bigger and bigger till by the time they reach the top they are even visible to the naked eye: little white curdled bits floating on the surface.

On this morning that takeover is still safely in the future. All that happens with this wondrous water-mineral-vitamin-protein-fat-and-bacteria fluid now is that a little bit gets poured in the coffee or tea. And it's doubtful if even that's noticed, for by this time the bread has been toasted, the jam and marmalade are out and, as always, there's even something in an attractive little tub next to them all, ready to spread.

Over the toast comes the margarine, nice gloppy melting chunks, spread fearlessly, applied thick, dripping in and demanding more. It used to be butter on the toast, but since all the warnings about cholesterol and heart attacks butter has been banned from this sensible household. Only this lighter, thinner, fresh and non-cardiac wholly vegetarian margarine is being used. Or rather, the users think it's lighter, thinner, fresher, etc. A look at the way it's manufactured would suggest otherwise.

Margarine is made from fat. It was first invented in response to an award offered by the French Emperor, Napoleon III, after the urban rebellions of 1848 to find a cheap source of fat for the working classes who could not afford butter. Today there's soya fat in margarine, also the fat you get from squished herrings, and about 20 per cent of the total is beef fat or even nice old-fashioned lard — pig's fat. All these fats are mixed together and melted, and if you think molten pig's fat smells bad, you should wait till you've had the misfortune to walk through a factory where it's being stirred in with boiling herring and other fats. The whole mess is so repulsive, so clearly distasteful and unmarketable (it comes out colored grey on top of it all) that before anything else is done it has to be funneled into even larger deodorizing vats to try to get rid of the stink.

What comes out of the deodorizer, while at least it can be approached without gagging, is still not quite the tempting substance commercial margarine is supposed to be. It's grey, it's sticky, and it's far, far too chunky. The fats that were temporarily boiled apart in the deodorizer can't restrain themselves for long, and have clotted back into large, unattractive lumps. Those lumps have to go.

The grey gloop is poured into another vat, metal shavings are clunked in before it, then the vat is screwed shut and high-pressure hydrogen gas is sprayed in. The fats are boiled and compressed in there, they react with the nickel and hydrogen, and when finally the ordeal is over and the top taken off, the lumps are gone, squeezed clear out in all the ruckus.

There's more. Beef dripping, lard and herring fat don't cost very much, but if at this stage they could be diluted with something even cheaper, even more easy to get in quantity than pig's fat, then the cost of producing the margarine would drop even lower. This extra substance is waiting in another vat in the factory, right next to the one where the de-lumping took place. It's milk — of a sort.

By government regulations there are two main grades of milk in most countries: Grade A, which is fresh, checked and suitable for drinking; and a lesser Grade B, which ordinary consumers don't usually see, but which being slightly older, or having a bit more bacteria in it than is best, is used for condensed milk, commercial cakes, and baby milk mixtures. The milk that's waiting to be mixed in with the fat in the margarine factory is the second kind, or even one grade below that. It's not fresh; in fact it's going sour. Even though it's already been pasteurized once, the factory engineers have to give it another pasteurizing heat treatment to get some of the worst stuff out of it. After that it's strained, filtered and poured in with the waiting fat.

This presents a problem. Oil and water don't mix (think of French salad dressing), and the lard and fish fat coming in through one spout are highly oily, while the sour milk pouring in from the other spout is that 88 per cent water. To ensure a match, other substances have to be sprayed into the vat where they're heading. Soap-like emulsifiers are squeezed in to foam around each drop of sour milk water and so keep it from joining up with the other milky water there in the vat. Then lots of starch is poured in to make the combined mix even more gooey, and to see that things don't slip back.

It seems it would take a genius to make a palatable product out of this soapy and starch-full mixture of grey sour milk and animal fat. Luckily, unsung geniuses aplenty labor in these margarine factories. First some color is added, something to cover over the vile grey. Normal yellow dyes wouldn't work, because the grey is so deep that it would keep on poking through. Extra-strong dyes based on sulfur-refined coal tars are used instead.

Then a stifling strong flavorer is mixed in to make it taste like something other than the miscellaneous lard, other fats and old milk it is. Then vitamins – because all this processing has made it nutritiously almost valueless. The result of all these labours is then compressed, cooled, scraped, cut into long blocks, cut into smaller blocks, and finally dropped into plastic tubs.

There's something else. Back at the beginning a little sunflower oil is sometimes mixed in. That's not because it's especially needed, and usually not because there's enough to make any difference, but rather so the designers can have something suggestive of sun-kissed meadows and open spaces to print on the cover. Even the French Academy of Sciences that sponsored the original stuff had their doubts. The award for the first margarine was granted in 1869, but eleven years later the Academy decreed that it couldn't be used in government cafeterias: it was, they said, too revolting on the palate.

With that toastly topping the breakfast preparations are over. Now it's time to consume, to chew, chomp, swallow and chew again. Only it must be done quickly, feverishly, for the clock in the kitchen is nearing eight and there is still much to do. The filling and warming substances enveloping the insides are one thing; now it's time to return to the bedroom, shake off the dressing gown, and do something about enveloping the outsides.

Fibers in a synthetic shirt are derived from oil-field petroleum, and squeezed out in narrow, identical strands. The red diamond in the middle is glue holding two such fibers together. As the wearer shifts and moves during the day he scrapes off the surface of these fibers, creating an electrostatic field that hauls in floating dirt

What to wear, what to wear? In certain provinces of Japan in the Edo era it was simple: dark drawstring trousers and jackets were apparently the rule, with exceptions punishable by prison. But in most other places it gets more complicated. There are denims ('de Nîmes' – the French town where sailors wore them) and there are pullovers, corduroys and flannels, khakis, kilts, ponchos, furs, full battle-dress, and even those little grassy jobs, woven together and held up with thongs, paste and a beguiling smile.

There has even been one more extraordinary contraption, a mixture of curiously slit trunk-surrounding uppers with an abdomen draping and limb-encircling bottom; a garment developed for hunting in certain forested regions off the coast of Europe and which only took up in final form about 1860 when it spread from the warrior fox hunters of the English aristocracy to the status-seeking professional classes of the same island's industrial cities, and thence around the world: the suit.

Its evolution has persisted. Perhaps half of the businessmen of the western world work in one of these equestrian-evolved get-ups, and although there has been some variation since the early days, some freedom of choice now available between under jackets and a single jacket, double cuff buttons and just a daring slit at the end of the sleeve, all suits the world over begin with one necessary preliminary, something to start the whole stack off as it were: the shirt.

Most are white, the next most poular seems to be blue, some are brown, hardly any are green or red, but all, without exception, are no sooner clambered into in the morning, no sooner stretched and pulled and held open and slithered on, than they start to get attacked. Dust and other particles in the air start swooping down on these initial torso wrapping protective garments, dust which gets caught between the fibers, or lodged in micro-pits within the fibers, or just lands gently on the surface of the fibers and sinks in.

This last-most adhesion is one of the most insidious. Even clean cotton shirts are dripping with all sorts of gooey outer layers on their fibers which will soak up such dirt. There are dissolved waxes, silica, resins, and even sterol plant hormones related to cholesterol and the sex hormone testosterone on every cotton shirt or blouse you've worn, while synthetic shirts, not content with their own sticky outer layer of residual finishings

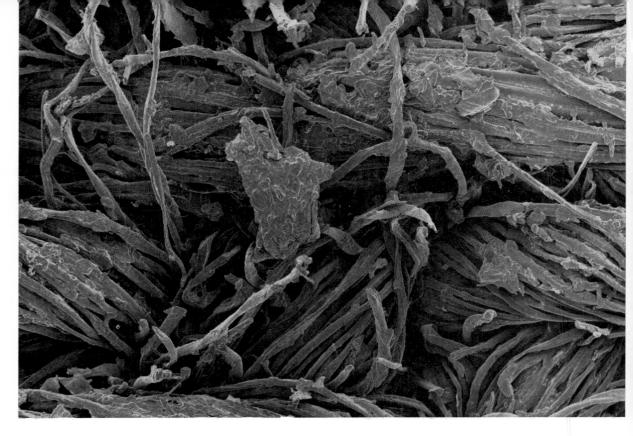

and plasticizers and hydrocarbon gums, have something else with which to pull in the omnipresent fine dust. They have electricity.

Rub one synthetic fiber against another of the same sort and electrons are ripped off from the friction. Rub synthetic fibers of different sorts against each other and even more electrons are ripped off – there's been even more friction. Every movement in a synthetic shirt starts some such electron stripping, and an electrostatic space-warp forms out from where it happened.

Any minute dust particle that has the misfortune to float by at just that moment is accelerated down into the warp without warning. Sometimes the particle doesn't even touch down on anything, but is just held quivering for hours in the gap between two fibers by the force field beaming out on it. Let these ambushes and space-warps repeat a few hundred million times through the day and a slight build-up of grime will be noticed. The shirt surface becomes a glued-on museum of all the chemical factories, car exhaust systems, agricultural fields and other sources of dust for hundreds of miles around.

Wearing synthetic shirts that are tightly cut provides some help. Then

Wear a cotton shirt for one day and this is what you get, magnified 160 times: dirty collar (above), clean collar (opposite). The spaghetti-like cords are what make up a single cotton thread. The gunk in the first picture is mostly skin flakes, skin grease, and errant aerial dirt — all held on by suction forces in the dead, but still potent, cotton

the electrostatic field wending out from the rubbing fibers will to some extent be counterbalanced by the identical, but oppositely charged, electrical field wending out from the continually friction-charged skin underneath. Loose synthetic pullovers get the worst of both worlds: no opposing field from the too far away skin; plenty of static build-up in the fibers to pull in the dirt. (Though to be fair, synthetic shirts worn this tight mean that a greater than normal amount of perspiration is likely to be sloshed onto the fibers, and this will hold onto nearby dust using the gooey surface mechanism we started with.)

Collars get an especially strong dose of grime from synthetics, as every tilt of the head will produce some rubbing up there, and so some entrapping electrostatic field. Cotton shirts don't have as much of this electrostatic boost but still manage to get dirty enough in the collar: as noted with dish towels all cotton fibers are hollow and filled with nice stretches of dessicated protoplasm. Dirt that's jammed into the cotton by simple impact can get attached to leakages from that protoplasm even if the cracks or goo on the surface of the fabric don't hold it.

If this sounds unfortunate for the suit-wearing member of the household, even more arcane indignities are being perpetrated on the nylon stockings that the other early-morning dresser is pulling on. There the fibers are not being loaded up with dirt — dirt won't penetrate nylon; they're not even suffering the bacterial build-up shirts are getting along with their dust — bacteria die from starvation on clean nylon stockings. What's happening is that sulfuric acid droplets are materializing out of the air and landing on the nylon fibers — and what sulfuric acid touches, it tries to dissolve.

The sub-visible droplets don't break through all at once. What they do is age the individual threads, break down the cross-weaves, and hasten all those other little snipping things that — though still unseen — will weaken the stockings enough for them to split in a giant run up the leg upon the first good scrape, or even, for no apparent reason, later in the day. But where in our clean house does the sulfuric acid come from?

Chemical factories and coal-fired generation stations will do the trick, but even if there are none in this neighborhood, none anywhere near the house, not for hundreds of miles around, there are still other possibilities. Volcanoes are an enormous source of sulfuric dioxide, which can float in the atmosphere for days till it gets caught in a cloud and reacts with the moisture there to come out again as sulfuric acid haze.

But again even if there are no volcanoes around, and not even any clouds overhead to invisibly trickle out the stuff that floated over from volcanoes further away, there is still another source of the acid that's landing in invisible micro-droplets on the stockings and starting to sizzle them away.

This is the lawn. Vast populations of bacteria living deep in the front lawn are always dissolving plant and insect remains that happen to penetrate down to their muddy lair. The nitrogen and other goodies in those remains they keep for themselves, but the sulfur that's in there they can live without. It rises back up through the soil, emerges around the grass blades and continues up into the sky. Sulfur from bacteria in other soil sources joins it, transforms into sulfuric acid in the air, and the whole finds its way into the house.

It will pass through the microscopic wall holes if there's no other way in. It goes through every room in the house and it will find the room with the stockings and when it comes within range it descends. Sulfuric acid

How nylon tights hold together. The individual nylon cords, (here magnified 400 times) are not glued, but knotted. Pull in one direction and the knot deforms and stretches a great way before it will rip

even at this dilute level will have some effect on eroding metal, paint and stone; the wonder is not that it attacks the stockings, but that they ever make it out of the house intact. The manufacturers give a little help — fillers, solvents, silicon and other defenses are mixed in with all nylon stockings — but considering that nylons would be almost immortal without the sulfuric acid attack their concern is perhaps kept within the limits of financial rectitude.

Back to the male. The shirt is finally on, and now only the problem of getting the trousers pulled up remains to be faced. The ancestor of trousers was the skirt, and that noble garment has always been easier to manoeuver. But a skirt that has been sewn into two narrow tubes, and is held together by an unwieldly saddle-shaped wedge of cloth at the top, is certain to demand yanking and tugging to get it in place. There is even, though the dresser doesn't know it yet, tearing.

Not great crackling rips, not the sort to make one spin around at parties and hastily seek the nearest sofa refuge to hide the damage. These universal early morning tears are smaller. The force lines produced when pulling trousers on work through the gridwork of woven fibres. If the trousers are of loose wool the force lines just shimmy about in there and fade harmlessly away. But if the trousers are of a tighter weave, if the pulling gets concentrated even for a moment at the indelicate seam holding the two legs together, then the force lines do something more before they disappear.

All their force homes in on just one thread — that's what a tight gridwork of fabric means. And the threads in the crotch are inherently weak, having been tugged around an industrial needle dozens of times in the sewing. The threads aren't even solid tubes, as they appear to us, but are woven braids of what seem like miniature piano wires. With this morning pull the first piano wire snaps and thrashes loose, then another does the same, and in the one-thirtieth or so of a second before the destruction stops the braid of fiber has begun to seriously unravel. An inopportune tug, bend, or even just good fidget later in the day and it can go.

Even with that avoided for now the freshly pulled-on leg wrappings still have to be closed. The great stuffed cotton wads under double-buttoned flaps that did such service as cod-pieces in more flamboyant ages are no longer considered proper. The simple zipper has taken its place, and although the wriggling effort needed to close it can induce even more thread ripping, its merits have caught on. Zippers are probably the most widely sold simple machines in the world, with production since their invention in 1891 estimated in the trillions.

With everything safely zipped there only remains the matter of walking. The pumping of the legs against the fabric sends incredibly dense clouds of skin flakes whooshing bellows-fashion out through the gaps between the fibers. Each step bellows out some more. The walking trouser-complete gentleman becomes a portable high-velocity skin-flake emitter, puffing them out in the thousands with every step. By the time he has reached the dresser for the tie that will complete his costume, a flock of invisible skin-composed cloud formations are left behind. They're so light — sinking at most one inch per hour – that many will still be floating when he returns at night.

One final preparation for the lady now, and then the dressing will be over. It is not a preparation handed down from time immemorial. Until quite recently respectable women did not wear make-up. Color on the face suggested passion, and passion was what they were supposed to avoid. Shortly after the First World War lipstick was referred to as only being appropriate 'to repair the ravages of time and disease on the complexion of coquettes'. They were probably the only ones to put up with it, too, as it was then little more than a greasy rouge, containing crushed and dried insect corpses for coloring, beeswax for stiffness, and olive oil to help it flow – this latter having the unfortunate tendency to go rancid several hours after use. The New York Board of Health considered banning lipstick in 1924, not because of what it might do to the women who wore it, but because of worry that it might poison the men who kissed the women who wore it.

For the liberated woman of today, the product has been transformed, re-thought, entirely remade. Insect corpses have been expunged as a barbarity; beeswax and olive oil have been rejected too. What goes in tubes of lipstick today is only what the best of late 20th-century cosmetic science can devise.

At the center of the modern lipstick is acid. Nothing else will burn a coloring sufficiently deeply into the lips for it to stay. The acid starts out orange, then sizzles into the living skin cells and transforms into a deep red where it sticks to them. Everything else in the lipstick is there just to get this acid in place.

First it has to be spread. Perhaps at some time you've noticed children playing with softened food shortening, smearing it over their faces. Such shortening (hydrogenated vegetable oil, as in Crisco) spreads very well, and accordingly is one of the substances found mixed in with almost all lipsticks on the market. Soap smears well too, and so some of that is added as well.

Unfortunately neither soap nor shortening are good at actually taking up the all-crucial acid that's needed to do the dyeing. Only one smearable substance will do this to any extent: castor oil. Good cheap castor oil, used in varnishes and laxatives, is one of the largest ingredients by bulk in every lipstick, from the finest French marks on down. The acid soaks into the castor oil, the castor oil spreads on the lips with the soap and shortening, and so, through this intermediary, the acid is carried where it needs to go.

If lipstick could be sold in modified shortening jars or castor oil bottles there would be no need for the next major ingredient. But the whims of the lip-conscious consumer do not allow for such ease of packaging; the mix has to be sold in another form. It must be transformed into a rigid, streamlined stick, and to do that nothing is better than heavy petroleum-based wax. Such wax can soak up shortening, soap, and acid-impregnated castor oil, and it will still have enough stability in its micro-crystalline structure to stand up firm. It's what provides the 'stick' in lipstick.

Of course certain precautions have to be taken in combining all these substances. If the user ever got a sniff of what was in there (all that castor oil) there might be some problems in continuing consumer acceptance. So a perfume is poured in at the manufacturing stage before all the oils have cooled – when the future cosmetic is still what the engineers call a 'molten lipstick mass'. At the same time, food preservatives are poured in the mass, because apart from smelling rather strongly the oil in there would go rancid (pace our ageing coquettes) without some protection.

All that's lacking now is the glisten. Women who smear on lipstick expect to get some glisten for their troubles, and their wishes do not go unheeded. When the preservatives and perfume are pouring, something

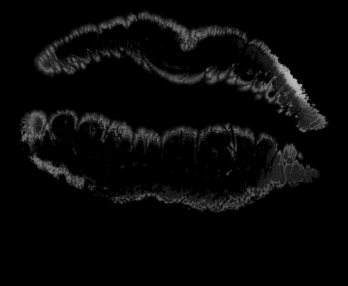

Lips, revealing
all the fine
contours into
which lipstick will
have to burn

shiny, colorful, almost iridescent – and happily enough not even too expensive – is added.

That something is fish scales. It's easily available from the leftovers of commercial fish packing stations. The scales are soaked in ammonia, then bunged in with everything else.

Is that it then? Shortening, soap, castor oil, petroleum wax, perfume, food preservatives and fish scales? Not entirely. There is still one thing missing: the color. The orange acid that burns into the lips only turns red on contact. That means another dye has to be added to the lipstick, a soothing and suggestive red one this time, so that what you see in the tube looks at least vaguely lip color and not a horrifying orange-juice orange. Which means, if you think about it, that the red dye you see in the tube has only a little to do with the color that's going to end up on the lips.

But such reflections are brought to a speedy halt – a glance at the clock shows it's late and getting later. The car is waiting, the office beckons, it's time to finish, to hurry, to speed. The house is about to be left alone – and the slamming door closes it tight.

Time's flow, caught in a 20 times close-up of the cranking hands of a wrist watch

TWO
MIDDAY

What's in the empty house, so abruptly left behind? A world of strange things getting settled for the long haul through the day.

When the human beings rushed out to the car their shoes scuffed through the outer layers of atoms on the carpet, like Dr Zhivago and Lara trudging through the snowdrifts, sending sheets of carpet atoms a hundred layers thick flying like so much powder as they kicked along. Some of the atoms whirled up high enough to land on the exiters' shirt and dress, peppering them with ultra-minute flakes of carpet orlon, polyester, and nylon; other of the drifting atoms stayed lower, and ended gusting in miniature blizzards against trousers, leg cuffs, high heels and stockings.

The humans who did this damage also received a 400-volt electrical charge in the process, but that will have discharged the moment they touched the metal doorknob. The carpet, left behind, has no such outlet, and is stuck.

For half an hour the electrostatic footprints carved out will remain clearly in place, marking the travellers' path, attesting to their way. (With the proper viewing scope the path would show a dim fluorescent green.) Only gradually do the footprints start filling in again, returning the carpet

Remnants and visitors in the house when we're gone. Left: The lingering presence of your heat image on a bed after you've got up. The upper image shows a naked subject on a bed; note the hot (white) hands and feet. The lower image is the empty bed, with the warmth pattern still visible to the infra-red sensor

Opposite: A fungus shortly after emerging from a spore, wafted along in the air. The red portion is the main body; the rising tube is the extending hyphae arm, which will dig through the plaster or brick of a house wall in search of food

Left: Electron microscope photo of a woodworm beetle, emerging from a chair leg

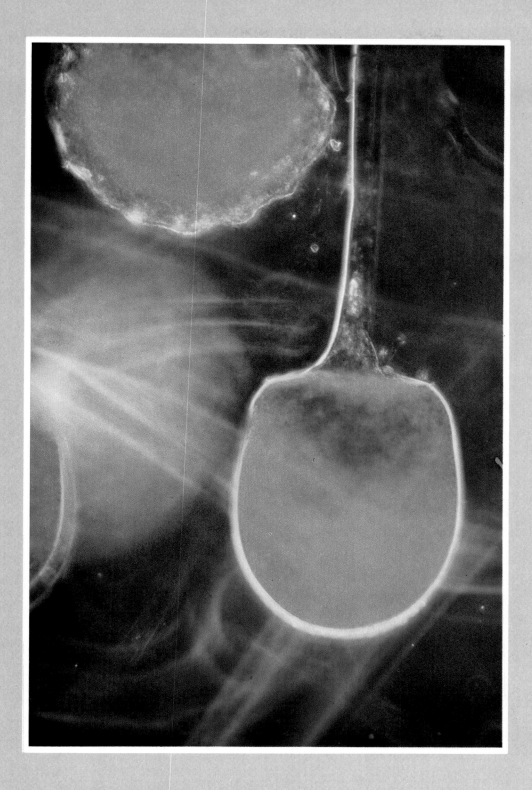

back to how it had been before being kicked apart. Some of the scuffed atoms on the surface pull down loose ions swirling through the room air; other of the atom clouds that were raised in the trekking settle down and add to the covering over. In two hours the tracks are quite gone, filled in and replaced, with only the memory of those who saw the journey out remaining to recall the way.

Walls are rarely walked on in the morning, but still suffer damage from the bustle before leaving. Water vapor starts it off, water that evaporated up from all the hand washing, teeth cleaning, face washing, face sweating, tea making, plate rinsing, table wiping and showering: 1½ pounds of the stuff on the average busy morning, all of which will have to come down.

If the air in the house were empty the water would simply soak into the wooden floorboards and make them swell, an effect of no particular significance during the day, and worthy of comment later only to those of delicate constitution who take the creaks and groans produced by their drying out during the night as a sign of some ghostly visitation. But empty the air is not. There are skin flakes in the air, shiny cadmium blobs in the air, textile bits and sea salt and lost micro insect limbs and also, hovering with all the rest, kept up by the pattering air molecules in every room and brought in by the suction pull from the slamming door, there are the fungus spores.

Spores are hardened containers which possess all the DNA instructions needed to create new fungus creatures, aerial eggs as it were. If they bump into dry walls they just rebound and go back to floating, but when they bump into wet walls they stick. Out of the broken open shell a single fungus creature's body appears, then a groping arm grows from that body, a leathery arm, albino and clear, and then from that arm grows another, and another, and then many, many others more. These are the fungal hyphae, and the reason they come out in such numbers, giving the appearance of something like a miniature and mutated octopus appearing from a micro egg on your room walls, is that the newly arrived fungi need them to feed with. For some species it's the sulfur grains in concrete that are sought, in others it's the metals in paint, or the glue in wallpaper, or even, for one especially abundant species, found at some time in almost every house in northern temperate climates, it will be the actual antibiotic poisons that the wood they land on produces which they slurp up and use as food. All over your house these freshly appearing fungus creatures will plug into the walls via their hyphae arm tubes.

Naturally not everything the arms absorb can be handled by the main body of the creature further back. Many are too poisonous, and any fungi that did accept them would lose its suction grip, unplug from the feeding spot, and fall poisoned to the floor. The way they get out of this problem is by spraying out the excess they don't need, releasing it in gaseous aerosols. At any moment during the day there will be freshly landed fungi on the room walls in your home, aerosoling out carbon dioxide, hydrogen cyanide, ethanol fumes, various alcohols, and much else. Concentrations are generally too low to detect except with special equipment, but there are exceptions. Some are favorable, as with those wall fungi related to the wild truffles, which produce an especially delectable emanation; but some of the exceptions are less pleasant, as with all the fungi that produce a musty or a rubbery smell. When the fungi levels are high enough to produce this you're likely to be able to see them too, enormous colonies, which appear to us as an unpleasant fur. But even when the density is too low to see they're there, many thousand separately plugged in fungus individuals on a clean home's inner walls. In older houses underlayers of paint surviving from the 1920s are likely to contain arsenic – it was used as an oil binder – and fungi plugging in from newer layers above will spray out an arsenic derivative like everything else. At one time this was a hazard, but now there's so little of this old paint left around that it doesn't matter. The last poisoning in England from an arsenic derivative released by wall fungi was in 1931.

As long as the condensing water is there for the fungi to emerge from aerial hibernation, there is no way to keep them from growing on your walls. Crawling out of the window so as not to open the door won't do much good, as windows are just as good an entrance point for the minute spores, and anyway some will already be inside the house. Most of the fungal spores you get were probably produced locally, and any especially damp house in the neighborhood will act as a fertile nursery source. Others come from further away, and the meteorological service's sampling aircraft have detected them arriving in Britain across the sea from all sides, with buoyant flights across the North Sea from Denmark taking only

Asbestos. It is a
stone that comes
in long threads:
excellent for
insulation yet
potentially
dangerous
when it flakes
loose

two days or so with a fair wind. In the US they can be blown from Texas to Minnesota within a week. As the spores can last up to forty years before hatching, you get your pick.

Aside from the fungi and footprints you might expect the house to just sit still and not make any trouble until you come home. But sitting still is just about the only thing the empty house does not do.

Some of the reason comes from outside. Sunlight is banging against the glass in the windows, and even though glass is quite impermeable to most solar rays, even though it's actually a liquid, continually molten, and only lets a very few of the sun's rays squish their way through, still there are such a large number piling up that even the small proportion that do succeed in getting across will do a lot. They heat up any table they meet, splitting formaldehyde loose from the varnish, and when they get to the floor they set that into motion too. The little hollows in carpet fibers heat up and begin a slow-motion, Medusa-style writhing in the light. It's not a total writhing, as wherever the light is blocked by the furniture there are infrared shadows that keep the carpet there cool, but it is enough to start the air currents going. The air rises up, slithering along the walls and bouncing off the ceiling. A few of the neatly stored piles of mite faeces in the carpets are light enough to be pulled up with it, but most are just heaved in slow motion to sprinkle from one spot on the floor to another.

Where the sunlight hits the walls it causes the heavier color-giving metal particles floating in the paint to vibrate like pinball flippers on the loose. But as the paint that seems to cover your walls actually works on an optical illusion, with large gaps between the metal particles which we only don't see because of limitations in our eyes, most of the light carries on through to the underlying material, the brick or wood or concrete or whatever is there. This stretches the material, it yanks it up vertically, it pulls on every nail and screw in the wall, and as the roof overhead is tugging upward from the sunlight too, the result is that the whole house begins to stretch. By the time you come home it will be several cubic inches larger — a solid addition that it will keep till night comes, when everything that was gained in the day sags away.

Even in rooms where there is no direct sunlight, there are still some curious goings on. Sweaters stacked next to each other in the bedroom chest of drawers leach molecules into each other in a process that is actually a slow dripping, while hangers in the closet are sagging from the weight of the clothes on them and giving off a detectable ultra low-frequency groan. On the dressing table top, rust is forming in thin water pools on a silver bracelet, atoms from a gold earring are bubbling loose and rising to the ceiling, while any string of pearls (made from caked mollusc mucus if natural, spun carbon if artificial) is crackling in tiny spherical bursts. A certain amount of carbon monoxide that was produced in the morning's cooking (carbon monoxide always emerges from the lighted gas flame) will combine with the water film on the silver and on the aluminum window fitting to produce dilute carbonic acid, similar to the bicarb in indigestion tablets.

From radium impurities in the clay, stone or wood of the bedroom walls a cloud of radioactive radon gas is wafting in, sparkling away in sub-visible flashes of antimatter after a few hours as it decays, while helicoptering through all the hubbub there are quite likely to be stony asbestos filaments, dropping down at regular intervals from the ceiling insulation, and speeded up by any vibration in the house's structure, such as the polyphonic chord sounds that a distant earthquake produces in the house's frame. (There are several hundred earthquakes every hour somewhere in the globe strong enough to make your house shake.) Even better are the jiggling and shaking of the house's full 200,000 pound, or more, bulk that arise when a lone truck drives past in the distance. In all this jouncing the hiss of the house's air escaping through the holes in the wall will likely be minimal, but there are many trillion microscopically quivering capillaries in even the thickest brick walls, and out the air goes, replaced by a new batch coming in through alternate holes every 90 minutes.

For hours the empty house's shaking, breathing, slithering and writhing can go on. But let the woman of the house slip home a bit early from work in the mid afternoon, and all these miracles of the living house are ignored. There's one thing she wants to do, and she won't notice anything till she has: she wants to get out of her clothes.

Office clothes are uncomfortable clothes, they cling, they catch, they tug, they inhibit. They're nasty bits of cloth we're forced to wear because other people think we should. If it's bad for us it was awful for the

Romans, who had to put up with great draping togas during the day, miserable garments which as Livy and Tertullian noted are hard to wrap on, impossible to keep wrapped on, and far, far too heavy to schlep around on a simple human frame all day. Claudius, then Domitian, decreed it obligatory to wear the toga garment; the Romans, proud heirs of an independent past, tried to resist, to shake it off, to replace it with tunic, or pallium, or hooded pallium; anything but more of the insufferable toga.

Our chosen replacement for official clothes is rarely the tunic or pallium, but rather of course the blue jean, a garment so soothing and suggestive of ease and mastery and laid-back lord of the manorship, which few of us bother to wonder about when we're tugging it on in relief, that it's interesting to speculate just how it was that this particular item came to be our standard of comfortable anti-formal wear. And indeed why it is that the color of this garment is blue.

It is a long history, but if any step had been left out the blue jeans would not be in this house at all. The story of the colorant in them will stand for the similarly intricate stories that could be assembled for the several hundred other important chemicals in the house.

In the beginning was the woad. This is a three-foot-high, woody shrub that grows in the forests of Northern Europe. Pluck off its leaves and dump them in something of a compost heap, and they begin to drip out a yellow juice. Rub that yellow on your clothes or your body and after a little exposure to air it changes color and starts to glow blue. You have made indigo — the most widespread blue dye used for over two dozen centuries of recorded history.

Druids used the blue from woad, and they were not alone. It crept into Roman underwear and barbarians' stockings; then in the Middle Ages it entered doublets, hose, frocks, wimples, jerkins and almost anything else that could use a colorful blue boost. The blue military uniforms that so many countries still use began then, because only that color, with its indigo secret, could be produced in inexpensive quantity. Indeed if your favorite color is blue, the choice quite likely stems from cultural habits begun in that epoch of the woad.

The history of blue dye since then has basically been the history of grown men acting like children. In the year 1200 English weavers started

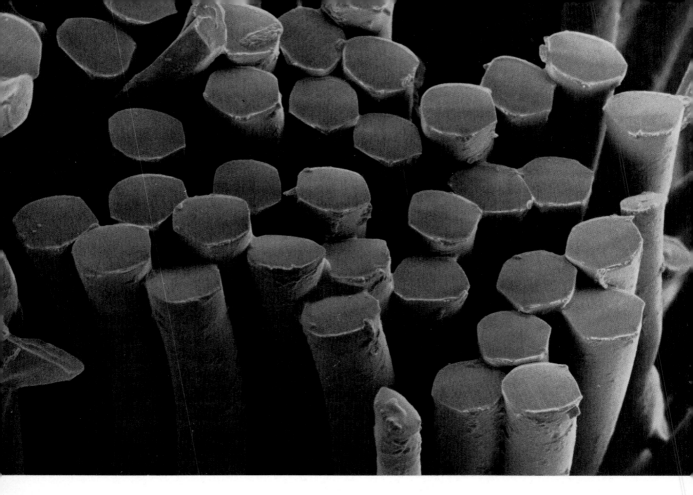

dying with woad (previously only dyers could). The official dyers retaliated and started weaving; weavers re-retaliated and refused to sell cloth to the weaving dyers. Trade stopped entirely and the country's blue faded until the King set things back to where they had been.

A few centuries later there was a bigger problem. With the terrific sales of woad-made blue in Europe, traders from other regions realized that if they had anything that could match that blue they would be in on a good thing. Dutch merchants travelling to the East knew where they could find that anything. In the moist sub-tropical parts of India there grows a plant that produces the same indigo blue dye as the woad shrub back in Europe. And being a sub-tropical plant it grows more quickly, and more cheaply, than the stunted European woad. Soon great quantities of low-cost Indian indigo began to arrive in Europe and especially Britain; just as soon great cries from British producers began to be heard demanding a tax on this foreign blue to keep it out. They requested the tax not because local wood

The truncated forest of a cotton-polyester fabric. The smooth tubes filling most of the picture are polyester, the middle ones hexagon-shaped by the pressure of their neighbors (as happens with honeycombs in a beehive). Wrinkled tubes leaning in from the sides are cotton

A velcro fastener about to close. Once the nylon hooks on top dig into the coiled surface below, nothing will be able to tear them loose — except a tug mighty enough to slice the hooks through the coils, producing that distinctive r-r-rip

production and their profits would be destroyed of course, but because of the health menace: the tropical stuff was solemnly analyzed and found to be (from a London document of 1577) 'harmful, balefully devouring, pernicious, deceitful, eating and corrosive'. The Royal Navy did its best to help, decreeing that its men could only wear uniforms dyed from good old local woad indigo, but that wasn't enough. The once-complaining British producers set up plantations in India and the Caribbean to harvest the variant sub-tropical indigo-producing plants for themselves. Luther preached that the decline in the woad trade was due to man's sinful ways, but his knowledge of relative costs was slight. By the early 1600s the last of the woad producers, left behind, went bust.

Those plantations dominated the market for centuries, until once again foreign competition began to foul things up. In 1885 a German chemist found a way to make indigo blue with just a few chemicals in his test tube. Simple once discovered, it had taken an especially persistent character to arrive at that simplicity: Adolf von Bayer was a chipper thirteen years old when he started trying to make artificial indigo, and a somewhat less sprightly sixty years old when at last a method that was practical succeeded. (Luckily he lived into his eighties to receive a Nobel prize for his labors; he also invented barbiturates, named he insisted, after his girlfriend of the time, Barbara.) Soon one German factory could produce as much indigo as 250,000 acres of sub-tropical British plantation. And there was more than one German factory. The plantation owners called for a tax on this pernicious synthetic indigo coming over the border. The Royal Navy did its best to help, decreeing that its men could only wear uniforms dyed from good old sub-tropical plantation indigo, but that wasn't enough. The once-complaining British plantation owners set up factories to harvest this variant synthetic indigo for themselves. By 1912 the last of the plantations, left behind, went bust.

At first the synthetic indigo blue had excellent sales – the lucratively uniform-mangling battles of the First and Second World Wars helped no end – but by the early 1950s there was a very big problem. The largest single indigo user in the world was the new state of communist China, where the obligatory uniform of workers' overalls for everyone was dyed with it. In 1953 though Mao declared that only local dye would be allowed, and so overnight 30 per cent of the world market for indigo was closed off. Against this collapse in orders even the Royal Navy looked unable to help. And, to make matters worse, new-style synthetic dyes, which could make lots of bright, inexpensive colors, were becoming available too. These had been held off the market for decades because the Swiss firm Ciba, which had exclusive patents on creating them, became deadlocked with the English multinational ICI, which had exclusive patents on transforming them to final product. Ciba could sue ICI to keep ICI from manufacturing it, and ICI could sue Ciba to keep Ciba from selling it. Only in the mid 1950s, just when the indigo manufacturers least needed it, did the two firms overstep their lawyers and enter into a joint-license whereby each shared in manufacturing and selling the new stuff. (The boom in bright and colorful cotton clothes in the 1960s was a consequence of this joint-licensing agreement.) Unless a way could be found to market vast quantities of the old indigo soon, a great many of the synthetic indigo factories would be forced out of business.

And then, some unsung genius, some shamefully uncelebrated chemical engineer, suggested dying trousers blue.

It didn't work. Who would want to go around in bright blue trousers? Nobody bought them. Indigo for trousers was a stupid idea, all the marketing people agreed. The engineer's genius remained unsung, his chemistry uncelebrated. By the early 1960s only four indigo factories were left in the world outside China, and they too would be closed down soon unless some miraculous new market could be found. It was at this point that another in-house chemist made an interesting observation. Cotton dyed entirely in indigo was too blue to be accepted as trousers. But if only *half* the threads were blue, if the vertical warp threads were soaked in indigo but the horizontal weft threads were left white, then something notably less garish would be produced. A small California textile company was found which had precisely this design on their books. It's name was Levi-Strauss, and its product was Levi jeans.

There's an interesting follow-up. The boom in Levis did not mean fresh indigo factories were started. The four old ones precariously left over from the 1950s near-collapse — one each in England, France, Germany and Japan — had been entirely written off against taxes, and in the case of those with creative accountants, had sometimes been written off against taxes several times. Their product accordingly was so cheap that no new factory could possibly compete with them. All through the blue jean boom since the mid-1960s every pair of jeans made has been dyed with indigo from one of those four ancient — the British one dates to 1908 — factories. American hippies and now English solicitors at home, Paris radicals and now Moscow upper-class teens — all have had their leg-coverings coated with this dye, chemically identical to the one the ancient Druids daubed for their arcane purposes from the leaves of the sacred woad.

Only with a change into comfortable jeans can our lady of the house move into the back garden to sit out and rest this afternoon, quiet and at peace until her husband gets home. Though what's going on under her feet is rather less at peace than she is.

Several feet under the lawn over which the woman reclines are a vast number of little holes. These are the pores, and in each one there are many creatures. The total amount of life is surprising: perhaps 10 pounds of living creatures – the equivalent of several slopping plates worth – will be active in these holes in the soil in an average-sized lawn. It's an excellent setting for life, there being abundant moisture, food debris tumbling down from above, and a nice constant temperature, moderated by the thickness of earth above.

There are two things these creatures do. First of all they murder each other, with the smallest bacteria in the pores being eaten by slightly larger protozoa, those in turn being eaten by the again slightly larger nematodes – vile-looking, pencil-shaped microscopic creatures with no eyes and six great lolling lips – and so on in a chain of six or seven creatures before it comes to an end. This murderous chain would not be especially interesting, except that to keep up their mutual assault the creatures in the holes under your lawn have to increase their respiration rate. They would run out of steam otherwise. In the process of breathing faster they incidentally dissolve certain sulfur and nitrogen compounds which the oxygen in our air has the unfortunate habit of getting stuck to; through their panting they release fumes which filter upward and indirectly but indispensably ensure that we poor surface dwellers overhead do not suffocate.

Along with this useful breathing, the creatures of our deep lawn holes are also busy synthesizing fluid dribbles that can be squirted out as defenses against other pore creatures that come too close. As such dribbles will be fatal to unwanted microbes on the surface too, it is from this source that we get most of our antibiotics. The satisfying smell of fresh soil the woman gets when sitting in the garden is due to the gases busily released by streptomycetes – the same creatures used to make the streptomycin and tetracycline antibiotics that hospitals stock.

This particular aspect of the soil dweller's battles incidentally provides one strong, if selfish, reason for preserving open land. Most of the antibiotics that we know today come from only a handful of soil creatures. The bacteria and actinomycetes are small, and their dribbles are smaller, which means that collecting them for tests is not easy. Probably 98 per cent of the sub-species living in our gardens' holes have not been fully tested. By past indications there's no doubt that many of them will have antibiotics that are as strong as our current ones, and probably some that are a good deal stronger. As many species live only in one small area,

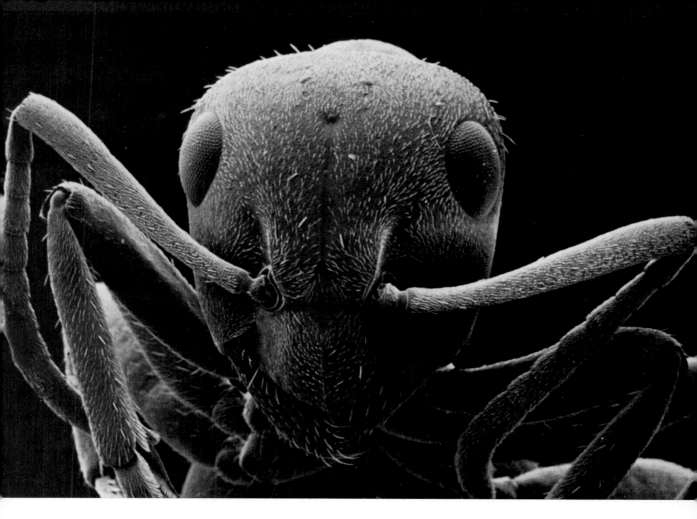

How a garden creature sees an ant. The antennae with ball and socket joint from mid-face stretch for details compound eyes would miss

every empty lot that is built over, every acre of farmland or forest that is turned into housing, makes it more likely that those life-saving antibiotics will never be found.

While the soil creatures are at work deep under the garden chair, certain equally curious activities are taking place on the surface all around the crossed feet of the woman resting so peacefully. There are the bumbling ants, locked into their hard armor and desperately trying to follow a scent trail that will take them back to their nest; there are the tiny click beetles, surveying the world in serenity from on top of the grass blades, but which tumble from their perch whenever an ant hits and then are forced to do dizzying triple somersaults to get back up; there are other insects, writhing on the surface from the after-effects of the phenol poison they swallowed when trying to nibble a leaf from a tree; but most of all there are the slime molds.

It is not the slime mold's fault that it has been saddled by science with a name not suggestive of the higher things in life. Under that unattractive label exists an exceptional beast. There are about 500 species, and we shall consider one of those most common in lawn soil, *Dictyostelium mucoroides*.

Most of the time this slime mold does not exist. If with the right instruments you take a close look at a portion of lawn next to your crossed feet where one might be going to appear, all you will see is a great number of microscopic amoeba creatures — shapeless bags of spreading protoplasm — squishing around among the grass roots. Left to themselves the amoebae would continue squishing around all day, but if you're so coarse as to jam your heel on the ground where they are milling, thereby destroying their food supply, then the individual amoebae will do a most peculiar thing. They will all stop what they were doing, stop still as if hearing a masterly call, and then, once the mysterious signal has been properly received, they will all turn towards one spot on the lawn, get themselves ready, and begin to crawl. The slime mold is about to form.

Not one amoeba, not a few dozen or even a few hundred, but thousands of millions, a number greater than the entire human population of the earth, all set out on this sudden and apparently senseless crawl. Some lose their protoplasm on the way, leaking it out through abrasive cuts opened in their surface and so die before they've made it all the way, but none of the other amoebae hurrying alongside them stop to pay attention. They're traveling in a fervor, a rush, what looks from above like frenzied religious migration, and do not let any misfortune that befalls one of their fellows shake them out of it. If there's a strong shadow from the lawn chair overhead they might swerve to stay in the light, but otherwise nothing will stop their rush towards the central spot.

For an hour the migration can last, the amoeba bumping into each other when they arrive to produce a ginormous pile-up, a miniature pyramid, rising in place at the center. This living pyramid is the key. What the amoebae are doing is building a structure from which a few of their kind will be able to flee the portion of lawn where they had been living till then. The reason is that with the food supplies threatened, or other stress, all the amoebae are liable to die unless something is done, and it's only with a fanatical migration like this that some of them at least

Pollen: bloated into a sphere to float in the air, produced in trillions which thwack into the nose with every breath taken in summer

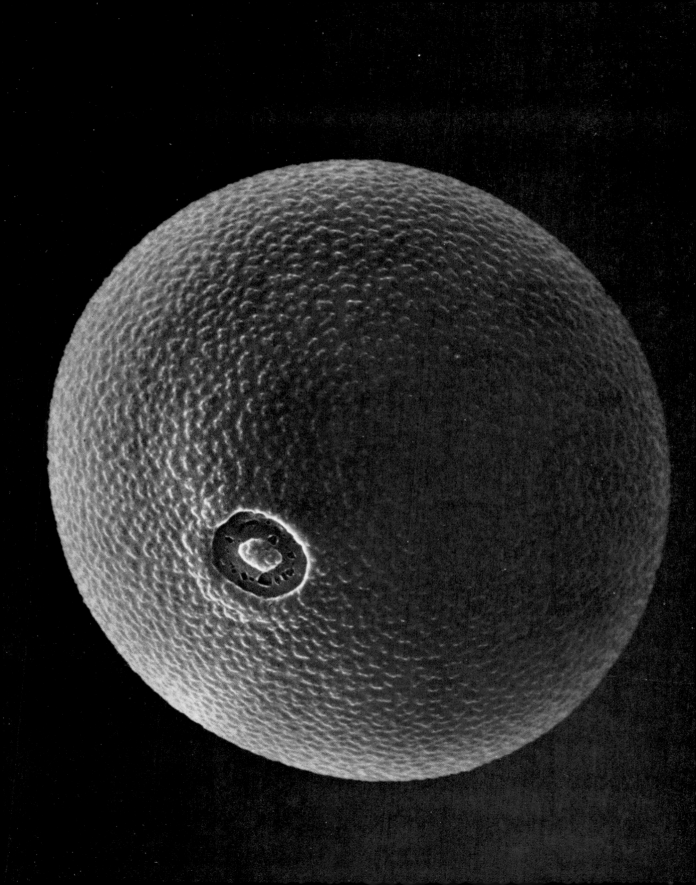

will have a chance to escape, and so carry the population's genetic inheritance on.

What happens next is that after a rest the pyramid of living amoebae begin to reform themselves into a tower. Since a tower shape is thinner than a pyramid, the same number of amoebae within it will rise higher into the air, and soon a minaret, a turret, soaring perhaps $\frac{1}{10}$mm straight up from the lawn surface would topple over unless it were strengthened, which is why the amoebae in its middle secrete a stiffening glue, a liquid wood, that hardens in just a few minutes to the same substance an oak tree uses for support. They die in the process, once-living bricks, but the tower is safe.

As soon as the tower is ready a certain very small number of the amoebae left alive within crawl up over the others to the top. There, in just another few minutes, they grow a hardened oval sphere around their bodies, an aerodynamic capsule sealing all the water and food they will need for a long journey inside, and then, thus ready, are heaved off the top of the tower.

The amoebae left alive behind in the tower quickly wither and die, all their energy spent in lifting off these few lucky individuals, the sole hope they have of carrying on their genetic material. Those travellers now have an aerial voyage of several hours or perhaps several months in store, flung around by the wind wherever it goes, quite likely whirling around the unsuspecting woman in the chair as they gain height, and then off to more distant realms.

A few imperfectly sealed spheres will come open in the air and the exposed amoebae cores will suffocate for lack of moisture; others will make it all the way to a landing spot and open as programmed, but find that they have touched down in some inhospitable region, an open cup of coffee perhaps or, worse, the parched world of the roof of a house. Yet all it takes is one to land in a spot where amoebae can grow, a spot where it can envelop bacteria and quickly divide to build up another amoeba population, and then the efforts of the now perished original amoebae will have been rewarded. It is an extraordinary escape story, a whole threatened population sacrificing itself to build a spacecraft for just a handful of their kind to escape — and it happens whenever you dangle your feet on the grass from your favorite garden chair.

In this backyard setting — the house writhing and breathing invisibly behind, the garden swarming invisibly below — the woman might well find herself succumbing to mid-afternoon hunger pangs and return to the home to retrieve a snack. For some it's marshmallow cookies, for some it's caramel-centered chocolates, but for our purposes we will assume her interest to be a good old-fashioned plastic-wrapped package of what in America is called chips, and what, confusingly, the British know as crisps. Whatever the name this is one of the most popular of all snack foods (165,000 tons of it sold annually in the UK), and a consideration of why it is so popular reveals much about the construction of a great many of those store-bought substances we put in our mouths.

First the potato chips bag has to be opened, and that's not easy. The plastic must be pulled at, tugged, strained — possibly hacked — before it splits apart. There's wrinkling, rustling, crackling; there's a straining face, taut neck tendons and, in the privacy of the garden, perhaps muttered but wild imprecations at the recalcitrant plastic before finally the sealant unseals, the plastic reveals, and with one crackling rip the contents inside come spilling out.

This regular plastic wrap battle is not due to some unfortunate oversight on the part of the manufacturers, some excess of sealing exuberance by a fanatic factory quality-control man. It's carefully designed in. Chips are an example of total destruction foods. The wild attack on the plastic wrap, the slashing and tearing you have to go through is exactly what the manufacturers wish. For the thing about crisp foods is that they're louder than non-crisp ones — think of chomping on a sausage as opposed to a carrot. The one squishes, while the other snaps. It's crisper. Destructo-packing sets a favorable mood.

Food rheologists — specialists in the arts and sciences of food crunchiness — have studied the subject carefully and come up with several necessary requirements for truly crispy foods. They must be loud of course, but mere loudness is not enough. Hot soup eaters or connoisseurs of soggy butter-soaked artichoke leaves are known to make a lot of noise, but no one would assert that the object of their affections is crispy. Crisp foods have to be loud in the upper register. They have to produce a high-frequency shattering; foods which generate low-frequency rumblings are crunchy, or slurpy, but not crisp.

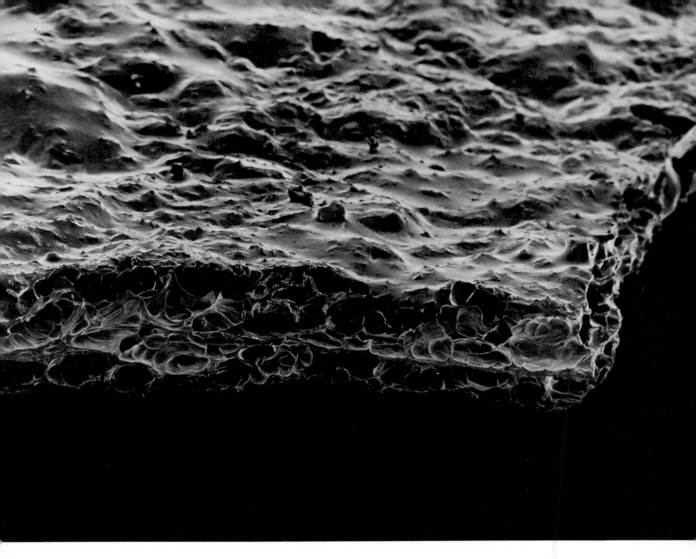

The things chip makers have done to ensure the auditory success of their products are manifold. The first gimmick is perhaps the most awesome. The chip you buy is too large to fit into the mouth (try it). This is simple, but insidious. A crispy potato chip must be snipped to a smaller shape by the front teeth before it can be placed in the mouth. Or if it does get crammed in whole, the mouth will have to be gaggingly wide open to accept it.

It's a shame we don't recognize this wilful choice of design. It is a work of genius. For a chewer to hear high-frequency crackles from what she eats her mouth must be open when she chews. That way sound waves can travel out of her mouth, around the side of her face and so up to the ears, without anything getting in their way.

Once the mouth is closed however, and the chosen substance is being crumpled away by the back teeth, there's no way any high-frequency sound components of that destruction will be able to reach the ear. Noise from this in-mouth chomping could only reach the ear by traveling directly through the jawbone and skull to hit it from the inside. That journey strips off any crispy crackle. Some high-frequency noise would be absorbed by soft tissues in the mouth at the start of the journey — tongue and gums especially soaking it up. The rest would get damped away in the skull bones, for the human head naturally vibrates only at the relatively deep 160 cycles per second — just about the E five-eighths of an octave below middle C. Only sounds in that bass range could get through, and such deep noises do not, as we've seen, a satisfyingly crispy food sensation make. That's why manufacturers make potato chips big.

Even with an open pathway to the ear ensured, there would be no tempting crackle if the potato chip just turned into mush upon being touched. It has to make a noise from the moment that it's first bitten. Few foods are so obliging. They thunk, squish, splat or crumble, but rarely will a food crackle. To design a properly noisy chip that people will buy, food rheologists have had to take their research lessons from those few members of the vegetable kingdom — lettuce, carrot, apple — that are naturally crispy when consumed. They make a crispy noise because they're composed of water-packed cells which pop open when you bite. When an apple or carrot is consumed, microscopic water jets geyser out at over 100 miles per hour at the point of tooth contact. How loud the explosive vegetable boom is will depend on how strong the balloon cell walls are, i.e. how much they can take before exploding. That's why fresh cauliflower is more embarrassing to gnaw in hushed public places than strawberries: the former has tougher cell walls than the latter, and so builds up more compressive force on the water inside before splintering. (Grapefruit, with its great sacks of water, comes with even softer walls, as those who have sat too near a spoon-wielding grapefruit addict in the morning can attest.)

These are the observations food engineers have incorporated in their design of the modern potato chip. Water itself is out, as it makes things that are going to be stored on the shelves perhaps several months before eating too soggy, but the key idea of little explosive cells is still in. The cells are just pumped full of air at the factory, instead of full of water, as at the plant roots. Any chip you buy is around 80 per cent air by volume. This allows for some interesting accounting. Air can be procured for nothing — just open the factory doors and in it comes — while once it's been repackaged in the form of tiny potato-wrapped pressurized cells, it can be resold at the price of chips. That's why large companies are so keen to get in on the chip manufacturing business, and spend so much on advertising to increase their market share once they're there. It is a very lucrative business.

Consider then the final consumption. The open mouth closes its incisors on the wonderful zero-cost air cells. It's not the walls breaking which produce the sound — those just rip in silence — but rather the remaining cell wall snapping back to shape which generates the air wave we hear. (A further delectable ping is added by any broken-loose fragments boomeranging at high speed inside the now vacated cells, like the lethal metal slivers broken loose inside an enemy tank by the latest shoulder-fired optically-tracked missiles, but that's a minor contribution.) The snap-back generates sound waves an octave or so above middle C as it starts, and as soon as it really gets underway, quivering and oscillating there at the broken-in edge, even higher-frequency harmonic sound vibrations are sprayed out in conical and expanding pressure waves from the point of the bite.

How to get sufficiently rigid cell walls to twang at these squeaking harmonics? Starch them. The starch granules in potatoes are identical to the starch in stiff shirt collars. That's why the chips we eat are made from a potato base. It's the extracted starch from the potato that's used. The whitewash which figures so large in Spanish villages and Mark Twain novels is near identical in chemical composition to the starch that gives stiffness to your chip.

However, starch alone will not do. Whitewash crumbles, and a chip made only of the stuff on Spanish village walls would crumble too. To get around that problem manufacturers are forced to add something else to their chips, a substance that is likely to be more abundant by weight than even the potato it is ostensibly made of. This addition is the fat. All chips are soaked in fat, great pools of often old stuff left over from other food manufacturing processes, before they're allowed to leave the factory.

The fat congeals, goes rigid instead of limp, and the indispensable stiffness is ensured. Over small regions it approaches the stiffness of concrete. A finished chip is 40 to 60 per cent congealed fat by weight when you bite it, but with sufficiently strong flavorers, and a certain discretion in advertising disclaimers, there's no need for that to interfere with sales.

So it's a shrapnel of flying starch and fat that produces the conical air-pressure wave when our determined chip-muncher finally gets to finish her chomp. Some of the sound vibrations at the center of the cone speed forth and are lost in the shrubs and trees on the far side of the garden, an unheard crackle of a thousand wasted cell destructions. But other sound vibrations in the cone take a different path, swirl back around the head, and reach the all-important ear without attenuation. The feedback that determines a properly crisp chip has been ensured. All it took was bubbles of air encased by starch and toughened fat – plus a geometrical masterpiece of audible food design.

What other artificial foods sound good? Just about half of all there are. Rice Crispies, flaked cereals, popping bubblegum, brittle chocolate, honeycombed candy bars, cheese crackers, chocolate covered brittle cookies: the list is very long. The Rice Crispy people even advertised for a while on the need to bend your head down and actually hold your ear over the ongoing milk/Crispy interaction to properly appreciate that product's true crispiness. Only what might be called the sensual mush types of junk food are free of the need for crackle production. These are the creamy, gooey foods, the over-sugared yogurts, the creams, whips, marshmallows, soft candy bars and their ilk.

This two-fold division is a good one, but where does it leave the poor manufacturers of *liquid* artificial food? It's easy to go for liquid sensual food – mush and crew are inherently wet – but how can you get liquid food that acts like crispy food? Crisp means having the mouth open for the noise to get out, and liquid foods, however well contrived, are sure to dribble, spurt or spray out of any test consumer's mouth who's being induced to hold it open to listen.

One solution would be a sort of miniature chin-tied bucket, like a nag's snuffle bag, to catch the overspill. But that shows limited thinking. Far better is to drop the attempt to imitate crispy or smooth, to forget these efforts to make a drink that gives pleasure either audible or tactile, but rather to go to the other side of the spectrum and instead pander to the perverse, the overlooked, the consumers who would be willing to purchase not packaged pleasure, but instead packaged pain.

That's what you're buying when you buy a carbonated drink, and that is what the woman was looking forward to when she brought back out into the garden a glass of Coca-Cola along with her bag of potato chips. All carbonated drinks rely on carbon dioxide dissolved in their water, and carbon dioxide works, as the medical texts remind us, because it is an 'excellent trigeminal [facial and tongue] nerve stimulant', attacking that nerve, as well as various structures in the soft tissues of the tongue, to 'elicit pain and a prickly sensation'. Its sub-clinical pain also increases saliva flow (as the experiment of alternatively providing such pain by lightly biting your own tongue will show), and that flow, combined with delight in the multitude of prickles, is apparently what constitutes refreshment – for enough people, at least, to produce the current carbonated drink sales figure of over 500,000,000 gallons a year in the UK alone. That's a lot of savored prickle.

The trick was conceived in the 1770s by the English chemist Joseph Priestly, who was the son of a Yorkshire cloth dresser, and who in those days before the professionalization of science was also a Congregationalist minister. This was useful. Churches and breweries were located near each other then – both being needed to cater to the varying needs of the human flock – and Priestly became intrigued with the curious bubbles that were produced in large quantity when malt was fermented for beer. Brewers usually vented those waste bubbles, but Priestly collected them, studied them, and even got the idea of forcing them, under pressure, into bottles of water. Then, brave soul, he tasted his concoction. The bubbles of course were what we call carbon dioxide, and the drink Priestly tasted was the first soda water. Luckily he was the sort to appreciate this abuse of the tongue, the trigeminal pain and prickle was pronounced good, and public sales of the newfounded medicinal tonic drink began.

The Coke you sit down with today is still basically Priestly's carbon dioxide in water mix. There have been only one or two small changes. When in 1888 a pharmacist in Atlanta, Georgia, bottled a variant of Priestly's water as something he called 'Coca-Cola', he was so proud of the medicinal pedigree that he marketed it as a mouthwash and gargle, guaranteed, in company advertising of the time, to 'whiten the teeth, cleanse the mouth, and cure tender and bleeding gums'. That was useful, but apparently of limited attraction. In time the gargle angle was dropped, and drinking the water, carbon dioxide, and sugar concoction was encouraged instead. There were still judicious amounts of cocaine poured in. To us that might seem excessive, especially as it was designed as a beverage for all the family, but this was before the first food and drug acts in the US. Brandy was common in children's tonic, morphine was available in many places without prescription, and by the end of the century a brand-new powder pick-me-up was being sold over the counter by an aspirin company: this was heroin. In that pharmacological flurry a little crushed coca leaf was not to be remarked upon. In 1903 it was dropped though, with labels of that date noting 'Cocaine Removed'.

Even without that extra something, cola-making became so profitable in the first years of this century – carbon dioxide costs almost as little as air, and that water to fill up the rest is cheaper even than the quite old fat of processed potato chips – that intense corporate minds were set to work to increase market share. A breakthrough came in 1916, a year when Europeans were distracted by the trivialities of World War I, but Coca-Cola bottlers at a convention in Terra Haute, Indiana, approved for some reason an unusual pinched-belly bottle design (the one still in use until recently). The reason was that this design, being so odd, could quickly be trademarked, and used as a legal device to bankrupt imitating manufacturers. As an in-house Pepsi-Cola history admits, 'Those were lean years'.

The woman in the garden is content. With Coke in hand, and open potato chip bag splayed not too messily in lap, the sensory enveloping that we like to create for our moments of relaxation is complete. Ears get the crisps crunching, mouth gets the bubbles exploding; even the nose is not left out, with errant drink bubbles rising upward to burst kamikaze style against it, and savory fat molecules from the broken crisp walls joining volatile artificial flavoring chemicals in evaporating up to satiate it further. Only one sound can disturb that suburban contentment: the sound of car wheels on the drive announcing that the other household member has arrived home.

The expanding bubble of a sound wave, pushing through the air at over 600 miles per hour, about to reach a waiting ear. Sound will carry slightly faster in a warm room than in a cool one

LATE AFTERNOON

THREE

The returning driver is in a skid. All cars are in skids, because all car tires melt as you drive on them. The rubber drips out into molten pools less than $\frac{1}{1,000}$ inch thick and the car slides on them. Evaporating as soon as the wheel leaves them behind, over 50 tons of rubber get into the air this way daily in a city the size of London or New York. The car is also shaking (enough to send spaghetti-like vibrations up the veins in the driver's arms), and in time with each shake is blasting out radiation from the jolting spark plugs, ungainly waves of the stuff, which whacks into the trees in the garden (where an inserted probe would light up to detect their arrival), bears down on the brass doorknob of the house's front door, the watch of a passerby two streets away, all other loose bits of metal in range, and, skeedaddling outwards as it does at 670 million miles per hour, even reaches the orbit of the moon a bare $1\frac{1}{3}$ seconds after release.

All of which is as nothing compared to what happens when the man enters the house to greet his spouse.

Normally when you talk, bubble-shaped sound waves come gurgling out from your throat, focus through the megaphone of the open mouth, and spread in ever-growing bubbles across the room. These bubbles are produced at a constant rate, and so follow each other in a stately procession as they float away. When most men speak the bubbles are pretty far apart, three feet or so between each one, and that's what registers as a relatively deep voice when the sound hits; when most women speak the bubbles press closer in on each other – a few inches less between these real floating cartoon spheres – and that's what is received as a woman's appropriately higher-pitched, feminine voice.

That's the theory, but it only works for stationary sound sources. Your stereo speakers, lacking legs, and rarely trying to escape on the pitiful stumps they are provided with, are standard examples. But humans, with their ability to jump, walk, twirl and back-pedal, make things more interesting. For a man stepping forward, bounding into the living room with a greeting for his spouse, the neat division between the sexes begins to go. For while his first expanding noise bubble comes out all right, the one after that isn't released where it should be, a good three feet behind, but instead is pushed up closer by the speaker's quick step forward.

It's the effect you notice when a police car screams towards you and the frequency of its siren seems to get higher and higher; here it's the hurtling police car that throws successive sound waves closer to each other than they should be. The effect is called a Doppler shift, fairly enough, after the Austrian physicist Christian Johann Doppler who first explained how it worked. Speeding towards you in a souped-up car, head leaning out of the window and passionate Italian opera emerging from his throat, even the deepest baritone of a lover would come out squeaking like Donald Duck. In the evening house the gentle stepping forward of the male returning home doesn't produce a voice rise quite so high, but it does squeak his words of affection just a bit out of their normal range.

This gender-bending sound bubble distortion does not work only one way. Just as the hurrying police car will suddenly drop in pitch once it's past you, so the squeaked-at woman in the sitting room will come out a noble bass, or at least lower than usual, if she's heading away from the man, perhaps trotting towards the TV to adjust it, as she answers. Her retreat pulls her voice air bubbles further apart from each other than they usually go, and that greater spacing is what we interpret as a lower frequency. Her voice has sunk, and the same will happen to any of us who speak in one direction while moving in the other.

What's delightful is that these effects go quite unnoticed by the person giving them out – *we* are always normal to ourselves – and it's in mutual total bafflement that the individuals on the receiving ends detect it. (The police officer has no impression that his siren is going up and down the scales.) At usual walking speeds in the house these register shifts are on the limit of ordinary hearing, but if properly recorded and enhanced by a Doppler analyzer they would reveal an extraordinary world: the burly male squeaking a falsetto greeting to his wife as his voice bubbles crash up against each other; the slender mate growling back deep ogreish responses over her shoulder as she steps away and her voice bubbles sink away from each other. If the wife were suddenly enamored of this meek-voiced hulk addressing her from the entry-way and felt compelled to rush forward and embrace him, a wild tropical beach imagined perhaps instead of a sitting room in deepest suburbia, her voice would rise back up from the deep, switch to its normal tone and continue up the scale to a bat-like squeak. Mingling with the husband's falsetto the enamored pair might sprint close, squeaking together for any observer with the proper equipment to detect, and perhaps maliciously later play back. The instant they reached each other though, relative motion stopped, comparative velocity nil, each one's voice bubbles would straighten out, line up back to normal, and so leave the embraced pair no longer warbling as two liberated soprano lovers, but back to the normal world of separate sex roles, and separate low and high pitched voices.

These frequency shifts are so precise that they were used to pinpoint distant aircraft in bombing runs at the end of the First World War. For a while in the 1930s it seemed that Britain would use this system as its main early warning air defense, and a few devices like giant ear trumpets were installed at intervals along the south coast, aimed at Germany, until the fortuitous development of radar led to them being hauled away and put in storage. That apparently immutable sound waves do slide about like this was first suggested by Doppler in 1842. He had no electrical equipment to measure the effect – there was none accurate enough at the time – so instead he arranged for a flatbed rail car carrying distinguished trumpet players from the Vienna Symphony Orchestra to go by while playing a constant note. Doppler calculated that the trumpet sound should drop a set interval for a stationary listener, and that is exactly what was detected by his pre-electronic receiving 'instrument' – yet another musician, with perfect pitch, who had been brought in and put in a chair by the train line.

A similar effect occurring in light waves rather than sound waves has been used to measure the speed that distant galaxies are travelling away from us. Working back it suggests that 18 billion years ago all the galaxies were on top of each other, and so it's by using the Doppler shift on display during every mobile utterance in your home that physicists conclude there was a Big Bang at that distant time to start the universe and send what would be the galaxies, including our own, bounding out.

After this interlude, the playfulness ends. The male has to retire to the kitchen, where duty in the form of an uncooked stew beckons, and the demands of dinner guests coming in one hour oblige. And while he goes, the woman decides she might as well watch TV.

What does her screen show? Dirt from Sweden of course. This does not mean updates of 'I am Curious, Yellow', but rather lower case dirt - soil, land, that sort of thing. It's mixed with other morsels of dirt from West Africa and spattered in a sticky mess over the inside of your television screen. When it's hit with electron signals from deeper inside the TV tube it glows, and as the Swedish dirt comes in different varieties – it's not a random shovelful you get on your tube, but carefully sieved first – the glow that comes out is in different colors too. The TV image you see is literally made of mud.

These colorful clods are known as phosphors, an interesting derivation from the Greek 'phosphoros', or light-bearing. Originally Phosphoros was the name of the morning star, which appears on the eastern horizon a little bit earlier than the sun. The name was dropped when the Greeks realized that the morning star was the same as the evening star, and in fact not a star at all but a planet. Their new label for this dual object was Aphrodite, after the goddess of love, whence the Roman and our own name for it of Venus. The original celestial label of phosphoros was eventually dropped, remaining only in a few chemical curios, the eternally viewed screen of the television among them.

The first of these rare earth phosphors was found outside the isolated hamlet of Ytterby, in Sweden, back in 1794. Explorers immediately set out to search for more. Some went in pairs, some went in great teams. One

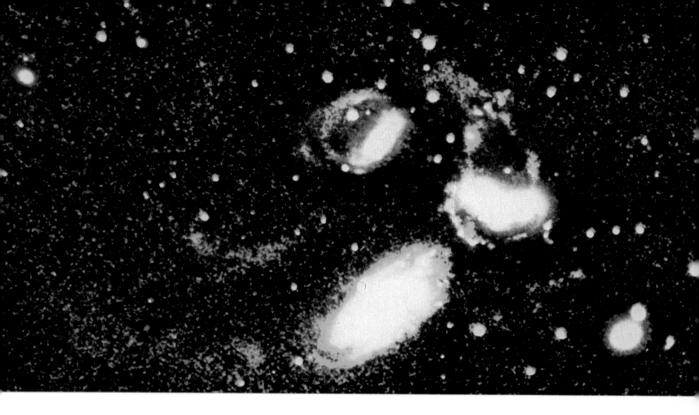

The Doppler effect in cosmic connection. Here are distant galaxies, each with thousands of millions of stars, quite possibly with inhabitable planets around them. Their light set off towards earth millions of years ago, and has been shifted towards the lower frequencies, exactly in accord with the Doppler relation that controls voice shifts

Russian researcher set out by himself, and after eight years' effort he found and isolated a second rare earth ore along a remote river. This poor man should have stayed comfortably at home in the warm, as not too long later the same ore was found in a slightly more convenient place; indeed the only spot on earth no one had thought of looking for it. It was found outside the hamlet of Ytterby itself!

Gradually scientists realized that the 'rare earths' were poorly named. They were not rare at all. Yttrium, for example, the first of them all to be found, and used now as the red phosphor in your color TV, is more abundant than lead. Nor are they specifically earths: traces of the same yttrium as in your TV have been measured on the moon and in the sun.

Not only are your favorite TV characters forced to go through their sit-com and other gyrations somewhere in this sheet of exotic mud, but to make the illusion of their presence even less believable is the fact that only portions of their body are actually projected onto the screen at any one time. The lighting up of the phosphor dots in time with the broadcast signals only happens in small, separated patches. The remainder of the screen is left jet black; we only don't notice it because this patch of temporary light streaks across the screen so quickly. Each square on your set is lit for only a few millionths of a second at a time. There's only nine millionths of a second for each square inch. Seen through extreme slow-motion filming your favorite show characters would be but horribly disfigured fragments strewn across the screen, as if the result of an industrial accident: there's J.R.'s left arm briefly visible over in the corner, then that disappears and Sue Ellen's twitching face comes on, lifeless, alone; each in isolation, glowing for an awful moment out of the void.

For the male member of the household over in the kitchen, sporting apron and with cookbook on French peasant cuisine laid out, things are not going so well. Where are the carrots?! They should be in the stew pot by now, there will be hell to catch if they're not, yet they're still sitting frozen in their plastic bag in the freezer. This is a catastrophe, and for extrication from food-timing catastrophes there is only one contrivance capable of rescuing the chef, one backstop, confessor, infallible make-doer, and that is the brushed metal box sitting over in the corner, a television with nothing in it: the microwave.

The carrots are put in on a plate, the door hurriedly shut, the on switch is pressed, and just sixty seconds later the transformation is made: what went in as orange icicles, good only for technicolor stabbings or brick wall piercing, will come out as pliable, temperate carrots, ready to be dropped in and help further succulate the stew. It's a miraculous transformation — especially as the microwaves never touched the carrot itself.

The door reveals what's going on. It's covered with a metal grid so that the largest individual holes you can see in through are fractions of an inch across. Cooking microwaves are too big to leak out and start doing their micro-work on you or any other already sufficiently well-fried object in range. They're not visible waves, but rather something like radar waves, only longer, and while they can't get out of the door of the microwave oven, they can reflect quite easily off the shiny metal plating on the wall inside it, and build up, a second or two after you turn on the appliance, to a density of many million per cubic inch. Anything in the way gets zapped, but since commercial microwaves are pretty gentle it would take a few hundred years for any solid objects to feel an effect from the great zap. Only if there's something other than solids in there can the microwaves get to work. That's why the plate holding the carrots stays cool — it's made of ceramic, which is solid — and that's why the main body of the carrot doesn't get directly affected by the microwaves either. But inside the carrot there is a lot of water (some frozen, much still liquid, curiously enough, as the carrot contains something like its own anti-freeze), which is just the right thing for the microwaves to work on with more success.

The water responds to the microwaves because water can be imagined as being made of chunks of oxygen atoms, each with two floppy hydrogen atoms hanging on like a beagle's ears. The floppy hydrogen ears flop even more than usual when the microwaves come on, and such flopping, along with the correlative scraping, grating, rubbing and gouging on the neighboring water molecules, creates friction. In just the way rubbing your hands together makes your palms hot, that friction sets the water to heating, and soon to boiling. This goes on in all the cells, and as the carrot (or potato or pea or other frozen vegetable) has millions of such cells, it's soon filled with millions of cell-sized tureens of boiling water. They thaw the freeze (they also wake up lots of bacterial spores on the surface that had been hibernating when they were cold), and produce the properly softened carrot that can save the stew, plus the chef's household reputation, from embarrassing ruin. Microwaves boil from the inside out.

We could suffer the same fate as carrots, as our planet in general, and our houses in particular, are part of a galaxy-wide microwave oven. But luckily the microwaves that bathe us, coming apparently from the rim of the universe and from the giant black hole that lives at the center of our Milky Way galaxy, pass no aluminum sheathing on their voyage to earth, and so don't get reflected and built up to dangerous intensity. All they do is make the water molecules on the chef's hands from the carrots, and the beads of sweat that surface on his brow from the stress, begin a gentle and quite invisible tremble as their loose hydrogen atoms flop ever so gently in response to this extrastellar call.

The microwave receiver used to discover the first extragalactic source, back in the late 1930s, was a giant of a device, about the size of a large house. Building a portable one was tempting, as it would allow a portable radar to be used. One of the first ever produced was taken from Britain to the US early in the Second World War, and kept, as priceless items are perhaps best kept, in a heap of newspapers under the British officer's hotel bed for several days while he tried, with increasing desperation, to find the right American liaison officer to take it from him. Since then the price has gone down, and the portable unit in the kitchen is one result.

The carrots are needed to join the meat already simmering in the stewpot. Such meat is rarely taken from the freezer, but was more likely bought fresh that afternoon, on the man's way home from work, a raw offering to the kitchen gods, a hunk of animal muscle to work by

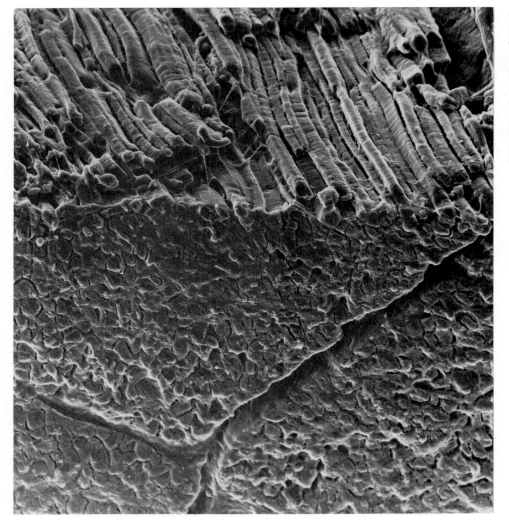

Cooked roast beef, showing the tube-like muscle fibers that in the live animal would quiver and get shorter when triggered by the brain, so making the animal move

sympathetic magic on the meek briefcase-toting bearer, and make him possessor of such raw animal muscles too. But was this symbolic muscle food fresh?

Certainly it looked that way at the supermarket, all red and glisteny and, well, *meaty*. But that was only its appearance, and such are known to deceive. Meat is red merely because the plastic wrap over it has been carefully built with lots of tiny gashes to let oxygen atoms down onto the haemoglobin on the meat surface. Oxygen plus haemoglobin comes out red, in beef as well as in our blood. It's so much like what goes on in a living organism that if someone puffed cigarette smoke over the display at the supermarket, the meat in its packages would quickly go grey as the carbon monoxide in the smoke poisoned the haemoglobin – just as within a cigarette-smoking human.

Ignore that red distraction and the truth is more easily faced. The prime cut meat you buy is over ten days old. This should not be taken as meaning that your supermarket is run by scoundrels and knaves who should be reported to the authorities. It's because you wouldn't care to buy your meat any sooner than this, not if you could get a look at it. Fresher meat is always suffering rigor mortis; it takes the ten or so days of waiting in storage for the tautened death agony of the animal's muscles to loosen out.

When a steer is slaughtered it doesn't all die at once. After the brain waves stop, continued haemorrhaging uses up the oxygen in the animal's system. This destroys the glycogen fuel in its muscles, and that produces lactic acid – the same chemical our unwilling chef gets in his legs when through a fate-tempting burst of activity, such as running after a bus, or playing squash at lunchtime with the club's resident killer, he uses up all the oxygen within. When a jogger says his legs feel dead, he's being more accurate than he thinks.

With all the lactic acid in our freshly killed steer clogging things up, the individual muscle chains that normally slide along each other can no longer move as easily as before. Little ratchet mechanisms emerge from many of the muscle strands, blindly and waveringly stretch out to adjacent strands, and then when they find them click into place and hold on. Tight. In a tired jogger these micro clicks and latchings are what appear as a general seizing up of the legs; in the more terminally tired steer the muscle ratchet latching gets tighter, and tighter, until it demonstrates what workers pulling slaughtered creatures from the ancient Roman amphitheatres called a 'rigid death', and which we, using their terminology without translation, continue to call *rigor* (rigid) *mortis* (death, as in mortician). In Raymond Chandler stories the descriptive 'stiff' is used, quite as appropriately. Only after 10-14 days of hanging will the ratchets let go, and so the muscles relax and go soft again. This is called decay. All red meat you eat has gone through rigor mortis and come out again on the other side, aged those necessary extra days for your delectation.

Chucked in the boiling pot there will be no more need to worry about that contribution to the evening's meal. The need will be to worry about everything else. The tasks start easy — cutting French bread, taking out the pastry flour bag — but then grow to fiendish complexity. Cookbooks give no help, being written only for experts apparently. One-off volunteers, especially male ones, are left to bumble, to guess, to make wild improvisations. In but a minute there are dropped pot-holders, maimed fingers, little clumps of congealed dough, and broken eggs with ungraspable slivers of shell within; in but half an hour there are *great* mounds of congealed dough, *lots* of opened eggs with ungraspable slivers of shell within, and also encrusted brown streaks from overflowed pots that have run down the stove.

But let the kitchen endeavor run for half an hour more, and there is a miracle: a finished two-course meal, standing amid the rubble on the stove and counter, there to be admired by its exhausted creator, its parent, its dad. What could possibly go wrong now?

It's probably best that the cook doesn't know. Into his masterpiece many things are falling. These include the usual items one can expect in a house, the perfume globules, hollow dust mite carcasses, stone asbestos fibers and other dust ingredients. A guest might be curious to find these in his first serving of the stew, and indeed a good peer with a microscope would show them all, but their presence is not very important. Perfume, mite mummies, asbestos, and the like do not grow. They're inert, and long since dead. What will count are all the living things that fall into the food. There are many of these around, and the simplest to start with is the creature shaped like a tiny submarine, oblong and streamlined, with a shell not of metal but of a semi-rigid slime, and which has 15,000 or so wriggling hairs extending from its body. Its name derives from the American veterinarian who first thoroughly described it, Daniel E. Salmon, and as the habit is to honor last names and not firsts, it is called not daniella, but salmonella.

Because these creatures are so small it is tempting to think that they're not really there, merely some sort of scientific construction. This is false. If you have got good eyes you should be able to make out a good range of dust flecks caught in a light beam in a darkened room. These can be as small as 20 microns ($20/1,000$ millimeter) long. Salmonella are about a tenth of that, so if you had only slightly better vision you could imagine seeing those hairy wriggling submarine life forms all about you. They exist not in a distant, unreal realm, but just a little beyond the normally visible.

The numerous salmonella now splashing around and exploring in the cooling stew most likely began as free-living creatures crawling on some exposed surface in the kitchen. They got there from any contaminated food that had been used in the past few days – commercial chickens are especially noted for being covered with them – and once released nestled in any moist place. The dish towel as usual is ideal, as too is the damp draining board next to the sink, and the incompletely wrung out sponge. Refrigerator door handles, regularly kept moist by the hands grasping tight to open them, are another good place.

Numerous other bacteria are prevalent in the kitchen of course. One recent study of several hundred clean, middle-class homes in the southeast of England showed the following percentage of homes that were contaminated with potentially dangerous bacteria of the kinds normally resident in the intestine: on the dish cloth – 97.8 per cent of homes; on hand towels – 98.8 per cent; sink taps – 94.2 per cent; the sink itself – 97 per cent; draining board – 99.5 per cent; washing machine – 89.5 per cent; refrigerator – 90.7 per cent; clearing up cloth – 100 per cent. The figures were as bad even in homes that used disinfectants, as most people had the curious habit of pouring disinfectants down the sink, which is an expensive way of using a drain, or of using the disinfectant *after* preparing food, when the damage has already been done, instead of before. Salmonella are just one type among these many, but since it is such an important one we shall just mention that these others exist, and continue to concentrate on it alone.

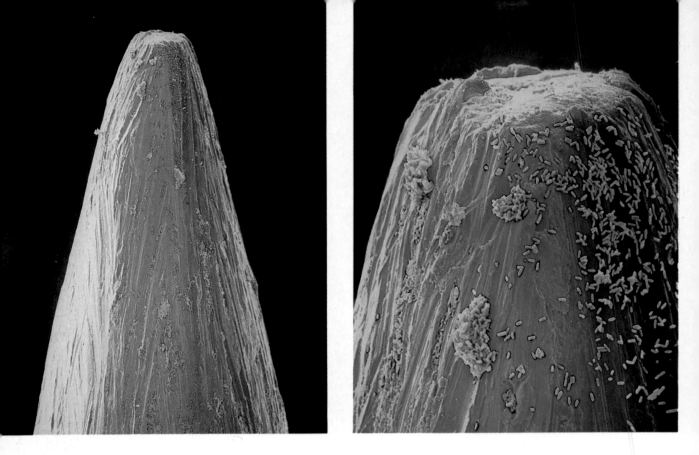

In all these kitchen habitats an accidentally transferred salmonella colony will rest, sitting quietly for several days, each passively soaking up the slight film of water about it, and each using its waving body hairs octopus-style to pull in the odd food granules. Without moisture the salmonella in your kitchen will die of thirst after a week, with only their gradually decomposing bodies to remain, but before that time any finger touching their resting place is liable to come up with what to us is a sub-visible micro-dot, but which for these creatures is a large family of several hundred members.

The head of a pin has numerous niches for life. Here are successive close-ups of a clean pin, revealing clusters of household bacteria

These bacteria are similar to some of the first life forms that developed on our planet. They have survived all that time because with their tiny size there are a large number of possible terrains for them to live in. Your house has as many niches for them as a new planet would have for humans – the vast Amazon of a wet sink basin, the quiet tundra of a dry hall table surface. (We will see later what appealing terrains the human body offers.) It is sometimes tempting to assign intentionality to these creatures, but that is misleading. They are only single, microscopic cells

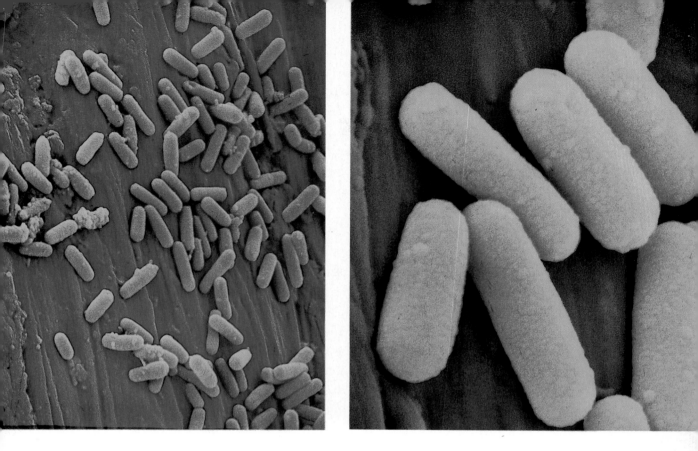

— mobile chemical factories as it were. There is no head, brain, nerves, id or superego: just life.

Once on your finger the salmonella try to adapt, to 'make themselves at home'. They're not disturbed by this sudden transfer to a living, moving object (except for the few that get splattered flat in the transfer of course), as they're far too small to take bearings and comprehend what was refrigerator door handle and what now is finger. In fact their first moments on the finger suggest that life is about to take a much improved turn. On the salmonella's scale a finger is a rough, swampy field. There are rolling hills which end in sudden overhanging caves. These are the top layer of skin cells. There are lots of comforting pools of slime and grease on the rugged surface too — these are the buildups of perspiration and other skin moisture, and the salmonella slop it up. That perspiration is nutritious: sweat on your finger contains potassium, sodium, zinc, glucose, vitamin C, riboflavin, and over a dozen amino acids. In this landscape the salmonella reproduce — either splitting in half, or gushing into each other long strands of DNA-containing protoplasm through holes in their body.

Yet in that setting of propagation and eating, all is not without its dark side. The fingertips that inadvertently sweep up the salmonella while hurrying around the kitchen are such a satisfying terrain that lots of other minute one-celled organisms have ended up on it over time. They react to the salmonella newcomers as a threat — that nutritious food perspiration comes in limited supplies — and set out as soon as they can to exterminate the new salmonella arrivals.

This microbial greeting party does not necessarily look a fearsome lot. All its members are blind, many are starving, and their top speed is limited to a slow crawl over the skinly bogs and swamps. When they get within reach of the salmonella newcomers, however, they show what they can do. Many spray out concentrated hose-streams of quite murderous antibiotics. A few especially emboldened ones quiver the short distance to the salmonella, and start absorbing all the food around, attempting, over time, to starve them out. The salmonella, taken by surprise, are slaughtered at first, but then recover. Some turn around and spray out their own brand of antibiotics against the attackers. (These defensive sprays kill not just individual micro creatures, but are pretty rough on the joined micro cell creatures that make up the human intestine, and are what produce the upsets of salmonella poisoning if any number get swallowed.) Other salmonella try to out-graze the closest attackers and send them away hungry; yet other salmonella, in the center of the colony, keep on with their reproductive swellings and squirtings, so producing more recruits to the threatened side.

These skin surface battles are going on all the time on the fingers of almost every one of us spending an hour in the kitchen preparing dinner. Rubbing the fingertips to the forehead in a moment of concern will spread a few of the salmonella, and so cause a new battle to rage there; drumming the fingers on the table top in impatience while waiting for the microwave timer to buzz will kill parts of colonies and totally upset the tactical arrangements, jamming the survivors deep into the bogs and swamps on the finger surface. Everything you do produces catastrophe for some, nirvana for others. An absent-minded scratch of the arm while puzzling over a cookbook recipe will land some on the especially moist and inviting terrain near the inside of the elbow; a subsequent touching of a finger to the tongue in preparation for turning a cookbook page once the recipe has been figured out will produce slaughter as the colonies are exposed to the fierce alkalinity of the saliva. There can even be moments of wondrous reversal. Opening the refrigerator for some extra butter is likely to drop some of the salmonella back where they started, the haughty attackers from the skin that get transferred with them now totally out-numbered by the hordes of salmonella on the fridge door handle.

And, mixed in with all the other actions, will be the ones that send the salmonella into the freshly prepared food. This might be a quick prod of the finger at a tender carrot slice that's poking out of the just turned-off stew, or a dab at the mashed potatoes. In the still quite hot stew almost all dropped-off salmonella will be sizzled and swiftly sent back to their maker. But in the rapidly cooling mashed potatoes, with plenty of good starch, water, and probably a knob of butter in there somewhere, they will thrive. They will thrive so well that every fifty minutes or so, with the usual solitary or group reproduction acts, there will be twice as many salmonella as before.

For preference the salmonella like to live on surfaces, close to the air they need, but in the mashed potatoes there is a good possibility that they will be able to expand into the deep, too. This is because of human fear of The Lump. Few hosts dare face the opprobrium that serving lumpy mashed potatoes to their guests would provoke, the sneers and raised eyebrows, the assertions that 'I *personally* never saw a problem with large lumps in mashed potatoes' from patronizing guests, the carrying of portable Black and Deckers with mashing attachments by the others. Accordingly the mashed bowlful is likely to be thoroughly pummeled, whisked, stirred, re-pummeled and, of course, mashed. In all that action the lumps are obliterated, but lots of minute air passages are created inward from the so thoroughly assaulted surface. Down into those cavities the salmonella will spread. It's dark, but the salmonella are sightless: the secret passages are just more places to copulate and increase numbers.

At this point, some perspective. The individual salmonella are each $1/10,000$ of an inch long. That means a plot of 900 in a square would still be a sub-visible dot. Nor are they the only microscopic creatures that have arrived on the cooling food. Other species come in on the fingers, fall from hair or beards shaken overhead while staring down – beards, being especially frizzy, provide excellent nesting spots for transported bacteria

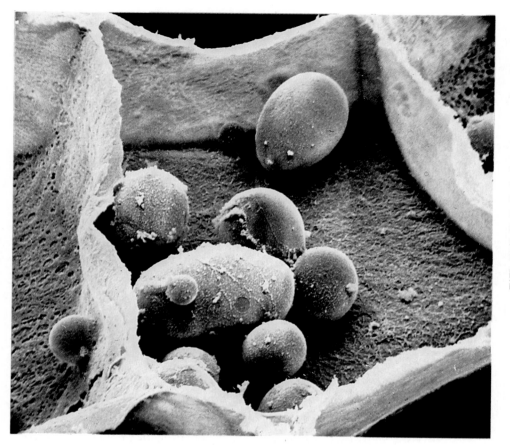

Inside a potato being cooked. Each chamber peels open, and the sticky starch (smooth oval blobs at this magnification) swells out. Digestive juices cannot peel open the cells, so without cooking the starch would stay locked inside, and the potato would be indigestible

— or arrive on that other near-universal source of drifting infection: broken off insect hairs that have floated in from outside. From all these sources perhaps fourteen different species, containing many hundred thousand individuals, are liable to land on the waiting dinner. Such numbers are, according to one report, what prompted Louis Pasteur, one of the first to discover these creatures on food, to carry a magnifier with him when he went to friends' homes for dinner, and whip it out before courses he had qualms about to study closer what was being offered. That was excessive of him. Almost all these species are innocuous, especially in the population of paltry hundred thousands you will get in the half hour or so food is cooling before being served. Only if the food is kept waiting hours will the numbers increase to an unpleasant level, whence the notorious reluctance of microbiologists to eat in cafeterias or fast-food restaurants.

Further entering this bestiary, if it's a summer evening and the window is open, there is quite likely to be another creature. It is a giant compared to the salmonella, and is even likely to be carrying a few thousand of those bacteria on its body, spilling them on to the kitchen floor, table, chef and other apparent furnishings as it passes overhead. It is one of the most agile flying beings in the animal kingdom, even if it was next to last in evolving this skill, following the beetle (300 million B.C.), pterosaur (150 million B.C.) and bird (130 million B.C.), and preceding only the bat (an *arriviste* 15 million B.C.). It carries on-board gyroscopes, has spring-release clasps to unhinge its wings, a catapulting undercarriage, and can even generate its own high-octane fuel. What makes it even more impressive is that this creature, *musca domestica*, commonly known as the housefly, does not look out of its swollen giant eyes at any world we would recognize.

For the fly sitting there on the kitchen table, reflexively stroking its front legs to clean them, and regurgitating some of the nice liquified dog excrement it lapped up earlier outside, the fluorescent light fixture in the kitchen is doing some very odd things. The bulb shines over the kitchen for a while as bulbs are supposed to, but then it suddenly cuts out, leaving the kitchen and everything in it in total darkness for a long interval until it just as suddenly lights back up again. To the human occupant of the kitchen, fussing over the food now being readied for display, no such stroboscopic flashes are taking place. The reason is that we can only tell two events apart if there's more than $\frac{1}{20}$ of a second between them. That's why movies are called movies, even though they're just a series of still shots projected a bit faster than that key $\frac{1}{20}$ second interval. A fly in a movie theatre would suffer no such delusion. Its nervous system operates so fast that it can detect distinct events happening only $\frac{1}{200}$ of a second apart. That would make a projection of Indiana Jones and his derring-do appear to be a tedious travelogue slide show, with long intervals of dark while the frames were advanced. The light in your kitchen goes on and off sixty times a second – that's how the power station pumps it in – whence the fly's moments of dark and light we started with. It's like what we might see in a curiously distorted disco, where the overhead strobes are slowly blazing on then off, and sometimes you see the other dancers caught in strange poses halfway between steps by the light, and sometimes you see

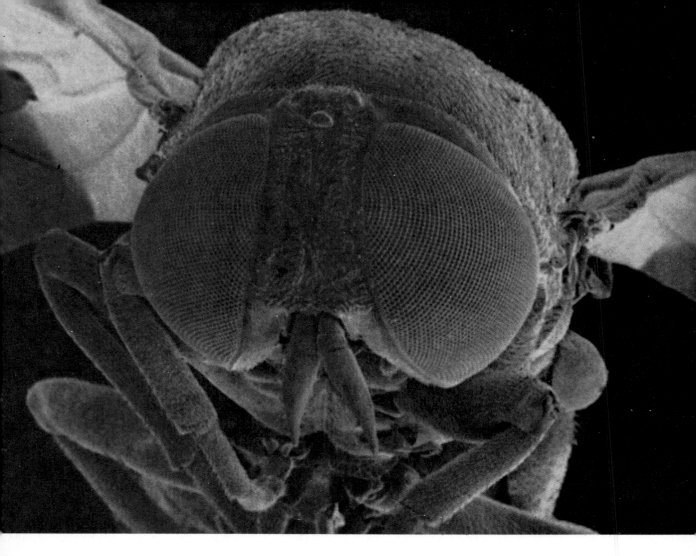

nothing at all. For the fly it's the final dragging of the stew tureen to a free counter that it sees in these disco stroboscopic blasts: strange and disturbing tableaux of grimacing face, slipping of pot-holder, nursing of burnt fingers, and soundless but deeply felt curses being emitted.

The fly would be content to stand on the table for ages, stroking and dribbling and watching this surreal slow motion show. But when the evening's chef spots the fly on the table, when he decides to take out all his aggrieved feelings at being forced into this infernal kitchen while his wife lounges watching TV in the next room – and during his favourite show! – when he decides to tear off his apron, symbol of oppression, and obliterate the cursed fly, then everything changes. Then the fly has to depart. It doesn't do this immediately, however. For the fly's vision,

How the housefly sees what we don't. The crystalline visors are actually its eyes, made up of thousands of small lenses that each produce a separate image. The hairless back aids streamlined flying though channeled through 4,000 crystalline little eyes, is not really good enough to make out fine details of movement over a distance. It's only when the human has stalked up quietly to the table, raised his hand and started bringing it down in a furious, murderous sweep that the fly gets a clear enough impression that its presence is no longer recommended. It can usually get away with its poor vision because it's fast. But can a fly really be fast enough when a first degree fist has already started on its way?

A human's hand swinging down at top speed will take at least $\frac{1}{60}$ of a second to cover the last three inches – the measured speed of a jab by Sugar Ray Leonard in his prime – and more likely $\frac{1}{30}$ of a second or even more. This the fly clearly sees coming, flashed into bright light by the slowly churning overhead fluorescent bulb, and without undue haste sets about preparing for take-off. A fighter pilot scrambling for his jet has nothing on the simple housefly, its technique evolved over 80 million years. First its brain works out the trigonometry of the descending hand so it can tell in which direction survival lies. Then starter muscles on the outside of its chest get their first signals to start pulling in the hard sheet of fiberglass-like material that forms the point of attachment for the wings. Those wing fasteners click inward and the wings are ready to be moved. That requires fuel of course, and so the fly – still in the interval while the furious hand is crashing the last two inches tableward – neatly opens its fuel valves. Not gasoline but high octane sugar streams into the muscles holding the wings, and great blasts of oxygen to help ignite it come pumping in through silvery air hoses. Only when the air and fuel are properly underway does the fly send a stronger hit to its starter muscles. The wings are pulled all the way down, like a propeller plane still on its chocks getting its first engine-spluttering rev.

There's no time for a running start, so the fly just tightens its thigh muscles, crouches slightly, then pushes straight up, thus catapulting itself into the air. The ungainly bug-eyed creature floats at first, like a helicopter hovering above an aircraft carrier deck, until its wings pick up enough speed to carry the whole weight. Then it turns sideways, retracts its landing-carriage legs to reduce air resistance, and accelerates fast, upward and away. The descending hand smashes onto the table, the burnt fingers get bruised again, and a strange cry of Anglo-Saxon origin sweeps out of the human's mouth. The sound catches up to the fly (sound traveling at 760 mph, the fly, though faster than us, doing no better than 25 or 30 mph), rocking it like a burst of air turbulence jolting a fighter, but otherwise doing no harm. By now the fly has likely had enough of this strange room, where men dressed in aprons come running at it with strange screams and lethal intent. It enters into steady flight, and sets course for the sanctity of the living room.

So far the fly has catapulted, evaded, controlled its fuel supply and accelerated away. It has never buzzed, and that's only fair. The buzz we hear from a fly is quite likely never heard by the creature itself. Flies flap their wings at about 300 times per second, and a sound with a basic frequency of 300 cycles is what we consider a standard mid-range sound. It buzzes. But to the fly, which samples and assesses simple incoming events ten times faster than us, that 300-cycle vibration is equivalent to only a 30-cycle fundamental tone. Now 30 cycles is nowhere near a standard piano or telephone buzzing sound. It's rather at the lower limits of our hearing, the sort of sound we hear when heavy machinery is clanking and rumbling nearby. If the fly makes the same transformation then its wings sound to it like heavy girders or boilers clanking slowly away too; the effect would be as if the fly's brain were that of a Victorian inventor, trapped inside one of his own flying contraptions that has suddenly started to work.

Because of its overspeeded vision the fly has the problem that half of its flying time – even with the kitchen light brightly on – is spent in what it perceives as utter darkness. And as any pilot knows, navigating in the dark is no easy task. (In the living room this won't be a problem, for the glowing filaments in the ordinary light bulbs there will keep light going even in the intervals when the power station cuts out.) The cruising fly manages to get through its cycles of night in the fluorescent-lit kitchen with the help of two gyrocompasses sticking out from behind its wings. If it accidentally yaws, pitches, rolls, or just gets dizzy when it can't see anything, the gyros inform the brain and a course correction is computed and sent along to the flight muscles. With all this help it can easily work out where it wants to go, and so the navigating fly cruises into the living room and ascends up to the ceiling, there to perform the most impressive feat of its journey: landing upside down.

LATE AFTERNOON

If a fly could travel upside down, landing on the ceiling would be easy. It would just have to put its feet out. But flies, like most airplanes, lose their lift when they try to go through the air bottom-side up, and become not flies, but sinks. How does a fly get around this problem? Watch one closely and you can see what it does. Proceeding at altitude high in the living room the fly lifts up two of its front legs as high as they will go in front of it. It's the position Superman takes when exiting phone booths, and it's ideal for what's to come. As soon as these two front legs contact the ceiling the fly will aerobatically tuck up the rest of its body and let momentum rotate it to the ceiling. The manoeuver leaves the fly's body suspended upside down, without it ever having had to do a full roll, a remarkable piece of topological extrication. From there it can vaguely detect the television screen flashing on and off below, and perhaps hope that the man in the half-discarded apron can't find him up there hugging the ceiling to attack again.

But in this household the fly is safe. No one's going to look for it right now. Everything stops, because man and woman have gone off to prepare themselves for dinner.

As the man is in the bedroom selecting his evening trousers, sadly, mournfully, accepting that his track suit bottoms with the comfortable elastic waist just will not do, the woman is in the adjoining bathroom taking care to scoop a mass of coagulated dirt, potato chip flavorant molecules, live fungi and other undesirables from a large webbing of dead cell material that she has hanging from her head. She is washing her hair.

What is hair? Dead cells, oozed out in rope form from numerous holes on the head. (There are about 170,000 holes and so that many hairs on a redhead's skull; the figure is 200,000 or so for dark-haired people, with blondes numerical in-betweens.) To lubricate the hair tubes as they press out from those holes, little sacks in the head squirt a molten lipid solution on each one. The common name for such molten lipid solution is grease. A little hair grease is a good thing – we would be pretty itchy if all our hair ropes were scraping out of all our holes without some lubrication – but too much hair grease and there's a problem. The molten grease gets colder as it soaks along to the further reaches of the hair, away from the warming scalp, and in cooling it solidifies, just as a drooping stick of butter does when given a stint in the fridge. Instead of useful lubricating grease you're now oozing with sheets of coagulating hard grease stuck to your hair.

With that wrappping, the individual hair strands are held stiffer than they would be otherwise. Rub a hand through them and they sway rather than bend: the hair, once flowing, is now greasy and stiff. That's bad, but there's worse. The grease packaging also picks up any of the things that happened to have been flying around up there since the last washing – the dust and dirt, the microscopic broken beetle fragments, the odor molecules (dangling nitrogen units from cut-open onions have an especial attraction to hair grease), the cigarette smoke residues, soot, textile sub-fibers, unfertilized pollen; all settle in, caught by the congealing natural goo. The total amount, even in a clean environment, is disturbing. Think of each hair as a slender strip of grease-smeared flypaper. Then the total length of flypaper we have hanging in ribbons from our head to catch passing substances can be easily computed. For a woman with an average nine-inch cut, that length multiplied by 90,000 hairs comes out to 810,000 inches, or 12.8 miles. That's a lot to wave through the atmosphere and expect to keep clean. (Even for a man with short and thinning hair the computation still comes out to several hundred yards.) Probably 15 grams of various aerial items will tumble onto the grease wrapping on a woman's hair and be caught since the time she washed it a day or two ago. In a year that is three kilos, or about a small bucket of slops worth; in two decades it's your own body weight, a hideously bespattered doppelganger, in unpleasant substances caught on the hairy head.

What to do with all the gunk? English nobles in the 17th and 18th centuries were assiduous in dusting cooking flour on to help sop up some of the grease, an ineffective technique which continues today in the coating on English judges' wigs. Marie-Antoinette took a different approach, and used piles of horsehair and gluey balls of flour in water to cover over what had accumulated on her head, and so keep up appearances for the ball. (Certain bread riots in the Paris region shortly before the Revolution are held to be due to the many tons of flour detoured to Versailles for the coiffures of young Marie and her ladies.)

Shampooing is an improvement on both methods. This is not because of all that luxuriant white foam: foam has no connection with cleaning, and

Not columns of plaster but human hair: solid tubes of dead cells squeezed out from numerous holes in the head. Left: Unwashed, showing dirt, skin flakes, odorous molecules and fungi embedded in the warm grease which slides along the hairs. Right: When washed, shampoo strips away all the undesirables

LATE AFTERNOON

Hairspray holds hairs down by coating them with liquid plastic, which then solidifies into rigid wrapping as here. In time, the hairspray will decay, cracking off into pieces which help feed the yeasts and bacteria living on the scalp

is put in as a separate ingredient entirely because some consumers won't buy shampoos unless they get their foam too. Rather it's because in shampooing there's detergent dissolved in the mix (shampoo is 15 per cent industrial detergent), and that detergent sinks down to the dirt embedded in the hair grease and plucks it clear. You can think of the detergent as a diver pulling handfuls of matted barnacles from a hull; in that case the foam is like the useless splashing on the water surface that might be produced by a pleasure boat overhead: if it can be seen on top it clearly can't be influencing the cleaning operation down below.

The detergent in your shampoo also tears out chunks of the congealed grease that trapped the dirt while it's at it, which seems good, but is in fact unfortunate. Anything strong enough to undo the grease is also going to be strong enough to undo the surface layers of the suddenly exposed naked hair itself. Hair so assaulted is a pretty pitiful thing. Normal hair is electrically neutral, with regions of positive charge balancing neatly alongside regions of negative charge. The shampoo pulls off only regions

Woman drying hair with hand-held blower, photographed with Schlieren technique that detects air-speed differences as multi-colored strands. Note direct heat blast on the left, which will dessicate and crack softened hair strands; whirlpools of heated air dance up with captured moisture from head

of positive charge, and so what's left behind is out of balance. It has a charge, each so-stripped hair developing perhaps a minute voltage during the operation, yet a noticeable total charge on the head when multiplied by all the hairs. These now-charged hairs crackle away from each other, writhing and twisting to stay out of contact, as if they were a field of pain-conscious prickly cactus terrified of swaying into each other, and so frantically trying to keep straight. Brushing or combing immediately afterwards makes it worse, for that strips off even more electrical charge. Brush 100 times, and it's approaching 100 times worse. Applying some refrigerated chicken fat or lard would solve the problem by imitating the original grease that kept the touchy electrical surfaces of the hirsute forest insulated from each other. But this secret tiptoe down to the kitchen is frowned upon, as is the well-pedigreed use of flour mentioned above.

That leaves only recourse to the other two masters of hair de-electrifying: the conditioner, and the hairspray. The former is just a mix that produces positive electric charges, to balance out the sadly isolated negative ones on the still wet head. It's hard to get the mixture just right, and this led to some problem with consumer acceptance when conditioners were first being sold. Those initial mixtures sometimes came unmixed in storage, and led to even more of the unwanted negative charges coming out of the bottle. Added to the negative electricity excess on the head already, the result was a double voltage 'Bride of Frankenstein' effect. Formulation chemists have since reformed and perfected their art; it's rarely necessary to go down to dinner with a paper bag over your head because of what the conditioner did. Hairspray ignores the electric mismatch behind unmanageable hair altogether. Actually just a kind of liquid plastic, it lands on the hair, forms into a continuous wrap, and clings tight. Nothing can get in or out.

Back to the man now, hurrying, impatient, eager to get into the till-now hogged washroom so his even more radical purification ritual can be carried out without more agonizing wait. But with which implement is he to perform the necessary actions? Razor or electric, electric or razor — the agony of the shaving market is that hirsute men refuse to decide between these two manners of slashing protuberant hair strands from their face. About half the male population goes for razor and foam, but the other half insists on electric and moistener. It seems to be a bit more for foam in Britain and America, more for electric in Denmark and Austria, but never, ever, does one sample manage to convince the other of the advantage of their chosen method and so produce the homogeneity that would simplify life for the manufacturers, who now have to keep two conflicting methods ready in production, enticing in advertising, and distributed to the stores. Worse, an estimated one-third of men are not even loyal to their selected hair slicing method, but instead waver from foam to electric and then back again: destroying production runs, confounding market share estimates, and generally driving the shaving executives who worry over these things nuts. Only older men show a satisfying stubble stubbornness: after the age of forty-five a hair extraction technique, having been chosen, is likely to be kept for life. And that is not because the older men are wiser, not because they have found a magic foam or ideal razor, but simply because they have become cynical before the tragedy of man's existence on the planet, and believe that whatever the ads say, whatever technology comes up with, hacking away very short hairs from their face is going to hurt.

They're right. Hair itself, as we have seen, is a simple tube of long since dead cells. The long ones on the scalp have been deceased for weeks or months, and even the short ones just edging out from the micro-pits in the chin are likely to have gone through the ultimate hair throes and died 15 or 20 hours before coming out. Cutting such dead tubes should be devoid of guilt and fear, as there is no living structure in them to carry a pain message and object. But it's different for the surrounding living face skin cells that have the misfortune to be located just next to where the dead tubes well out. On the microscopic level where these skin cells live the razor blade in an electric shaver, however sharp it looks to us, takes on the appearance of a jagged and rusty giant metal rake. Pull it across the skin in even the most painstaking of even passages and on the micro-level it will drag and skid, bounce up and then slash down, doing awful unmentionable things to the live skin cells in its path as it lands. Even when the blade does reach the sought after hair the razor's ragged impact is often so unsuccessful, so unlike the clean scalpel swipe we imagine, that the hair is merely gashed and will have to be tugged upwards to get loose, as a sapling that you have given one hack to with an axe and then got your blade caught in will have to be pulled bodily upwards with its roots. Microscopic analysis of the debris around a shaving barber's chair has revealed the sorry story: cutting off short face hairs means hacking away 'large amounts of varied epithelial components' – skin – along with them. The technical term for this is 'skin trauma', and it is a process that can't help but hurt.

Philosophical sorts might take consolation from the fact that it was worse for earlier generations. When Alexander the Great insisted that his soldiers shave so the enemy couldn't grab them by the beard, the best his supply officers could offer were rough bronze knife blades, or even filed flints, to do the job. Citizens of Imperial Rome even at the height of its power seem to have never heard of face oil or lubricant, whence the lament of Martial, possibly not exaggerating for once, at 'these scars, whate'er they are thou numberest on my chin, scars such as are fixed on some time-worn boxer's face', scars worse even than 'a wife formidable with wrathful talons (could) wrought'. When the newfangled straight-edge razor was popularized by the French King Louis XI in the 1400s to improve shaving ease, it was not long before wincing users had to rename

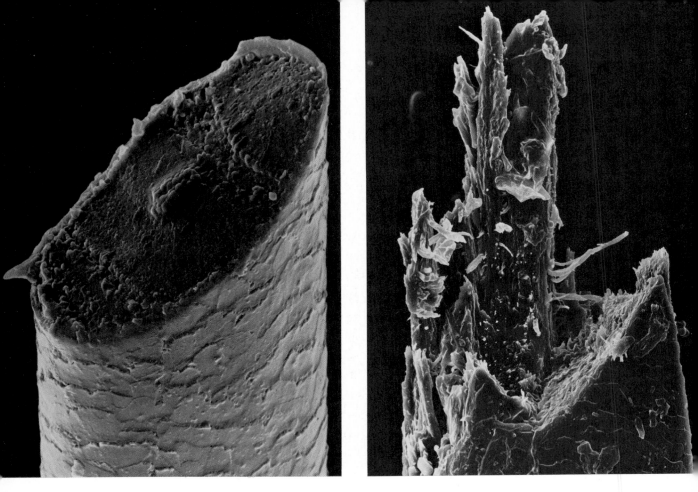

it the cut-throat shaver. Even the etymology is painful to recount, 'razor' coming via the mediaeval French from the Latin root *rader*, which means, horribly, 'to scrape'.

For the descendant of that late medieval cut-throat instrument, the stainless steel safety razor of today, the initial preparation applied on the face to ward off the worst is almost always aerosol foam. Shaving foam would be a wonderful thing if it could remove the layer of congealed grease plus embedded dirt which is wrapped around each hair of the emergent beard. Without them to go through, the scraping razor would have a better chance of severing the bristles straight out, instead of just uselessly hacking. But foam does not remove grease. Soap does, and indeed the old-fashioned shaving bowl mixtures one can still sometimes see at the barber's are based almost entirely on soap. But soap does not blend well with aerosol foam, so shavers who use the foaming aerosols as recommended in all the ads have to go through the shaving ordeal without

What the electric razor companies do not like you to see. Left: Beard hair, cut with an ordinary razor blade. Right: Hair from same man, but trimmed with an electric razor it. All they get is a lot of bubbled air (the foam), some petroleum wax and fragmented algae bodies to coat the skin from the worst of the abrasion, and, of all things, chemicals which give the face hairs temporary little erections before the shave, swelling the *arrectus pilorum* muscles at their base so they poke out an extra $\frac{1}{60}$ inch and can then be more easily truncated. It's not as good as soap, but cheaper to produce (all that zero-cost air again, with algae coming cheap too) and a lot more lucrative to sell. As an added boost, shaving foam is pre-mixed with hairspray, so it will hold firm while it waits on the jaw.

The aftermath of shaving is disconcerting. There are ripped hairs, gashed hairs, shattered hairs, and dangling hairs; the grease that had been peacefully congealing is sent flying all over the place, with dust and grit and insect particles that had been embedded in it flying all over the place alongside. On the razor or electric blade there will be 100,000 to half a million skin cell chunks that have been torn loose, and on the face left behind — fortunately at a scale quite invisible to the naked eye — there are dangling stumps, craters and gashes, now filling up with slowly welling blood from the finest — and again quite invisible — capillaries just below.

This is the scene of wreckage upon which the aftershave lotion is slopped. Manufacturers call them sprightly or bracing. This is an understatement. Almost all commercial aftershaves are 40-60 per cent pure ethyl alcohol. Think of how pouring 80 proof rum on thousands of tiny scratches would feel. With this alcohol the living skin that remains around the hair stubble will be seen through the microscope to writhe and jerk and shake. That flings some of the still partially rooted dangling hair stumps out; it also, and this is what it's there for, closes a good many of the microscopic cuts from the sheer shock of it all. The technical word for this effect is astringency — a derivation from the mediaeval French for 'to bind tight', and in fact from the same Greek root, *strangalan*, as our word 'strangle'. (The Romans, lacking our inexpensive purified alcohols, used vinegar-soaked spiders' webs.) An anaesthetic has to be added to aftershave to make this micro scratch strangling tolerable, and concentrated menthol oil plus, often, pure adrenalin is preferred. Its presence can be noted both by the smell of the menthol, and by the fact that any inadvertent splashing of aftershave on the lips is likely to produce a sudden and woolly numbing there – just as after an anaesthetic injection at the dentist. There's also a good deal of antiseptic mixed in the aftershave bottle, there to temporarily slaughter the skin bacteria that would otherwise take the opportunity of those thousands of razor-opened holes on the skin to escape down inside. Some colorant, some perfume (without it we just have the smells of a high proof alcohol mixed with disinfectant), and then the aftershave is complete. A few hours wait and the new layers of facial skin, pushing up from below the wreckage, won't know what happened.

If mutulation is more than wanted, there is always the solution the male Romans finally took. They endured daily shaving with painfully uneven hard iron blades for almost two centuries, from the time of Julius Caesar (who is recorded plucking his beard, too) to Trajan. It was only Trajan's successor, the Emperor Hadrian, who found a magical way to make the once uneven blades hurt far, far less. He grew a beard – and encouraged everyone else to join him.

While the man is now hogging the washroom, grimacing, cowering and yelping through his shave, the woman is back in the bedroom steaming. She is not steaming with rage, but with smell: from the female human body, recent washing notwithstanding, there wafts every minute notable quantities of ammonia, ethyl alcohol, acetic acid (the stuff in vinegar), hydrogen sulphide (the stuff in rotting eggs) and, above all, the dreaded mercaptans – the active ingredient of a skunk's smell. Most of these smell molecules crumble in the air after 15 minutes or so, but there's always more behind coming out, so the enveloping odoriferous cloud is constantly being kept topped up. Men of course steam with the same substances too, but they're allowed to go around smelling like that. Women, being theoretically composed of sweeter stuff, are not. Only one substance can possibly save the day, one mixture that will change her emanations from what medieval German peasants uncharitably called *stinkon*, to the more alluring mix which enamored Roman courtiers noted was produced with smoke, or *per fumar* – our 'perfume'. Originally orange-scented to mask faecal odors, in the days before regular clothes changing became popular, they have been refined considerably since.

Glop on some perfume to counter the natural bio-effluents and curious things begin to happen. Perfume in the bottle is 98 per cent water and alcohol, 1.99 per cent grease, and, the leftover, 0.01 per cent perfume molecules. Those three ingredients are not randomly mixed. The grease forms into little blobs and floats at various depths within the water, while the perfume molecules separate into smaller bits and sit on top of the grease blobs. Lifting the perfume bottle up from the table will shake apart these grease blobs and dump the perfume molecules off into the water, but in the time it takes you to unscrew the top, as long as the bottle is held at all steady, the grease balls will re-form, and the perfume molecules will speed back to their resting place on top of them, like sleek seals wriggling their way to bask on their favorite colored giant beach balls. Poured on your skin the 98 per cent water forms very shallow but quite pure lakes; the grease spheres tumbling out of the bottle with them land on the several thousandths of an inch thick seas so produced, and bob gently up and down. The tiny perfume molecules, quite unaware of what's going on of course, just continue resting on top of the grease balls.

So far the scene is no different from what might happen in some sort of gargantuan earth-side circus. But since the perfume and grease balls we're interested in exist on an exceptionally reduced micro-level, it's not long before the differences begin. Think of the perfume molecules as seals again. On earth a seal on a beach ball might kick its tail up, or even briefly toss its whole body up, but this would be rare, and it certainly wouldn't take long for it to fall back down again. Not so the perfume. The little molecules resting on the grease balls spontaneously unhook themselves and float up a few times their height above the water. They don't weigh enough to come down, so there they hover. This alone is curious, but there's more. Even if a normal circus seal could be induced to float upward, perhaps having been inflated with great quantities of helium gas before being sent out, it would be unlikely to do anything else more than just wobble in place up there a bit. Again it's not so for the immensely smaller perfume. On the size scale of a hovering perfume molecule the normal atmosphere in the house, invisible to us, is a maelstrom of large flying pellets. These pellets are the air molecules, and there are so many of them, belting in from so many sides and so fast, that the hovering perfume molecule is knocked higher and higher. A hovering seal bombarded with thousands of discarded tin cans and popcorn boxes from disgruntled fans in the cheaper seats might suffer the same effect, but for the perfume molecule the incoming pellets are numbered not in the thousands but in the many, many billions. Its departure is very rapid indeed, with individual perfume molecules being propelled to over 180 miles an hour from the impacts. That's what will whap into the man's nose when he sniffs. The bobbing grease ball becomes a dot quickly receding from view.

If a storm is approaching the house there will be less air in the bedroom (that's what an approaching low pressure system means), the perfume will have less blocking its way once it gets going, and the grease balls will recede below even faster. Such faster moving scent will reach waiting noses more quickly. Daub on perfume on a stormy night and it will smell more strongly, and be noticed faster, than on a sunny day. The effect is even stronger in high altitude cities like Denver or La Paz, where the air pressure is always low and so any perfume once applied will be even more quickly spread out. Naturally, the more energetic perfume molecules — they all have a certain amount of quivering vibration — are the ones that will get to take off first, and that means the ones left behind after the first wave of departure will be preferentially the ones that by chance were cooler, more sluggish, than the rest. That could mean the end of the perfume's favorable wafting effect long before the dinner guests even arrive, except for the precaution of pouring the stuff on those points at wrist and neck where the body's arteries come close to the surface. Blood in those arteries is carried at blood-warm 98.6 F, so any perfume poured atop gets heated for free by those giant heating coils down below. The shallow lake of water warms up, the bobbing grease balls on top heat too, and the sluggish leftover perfume molecules resting on the grease are stirred from their lethargy and given the heat boost that will send them floating then jetting up too.

Equally odd on the micro-level is the anti-perspirant the man is engrossed with while the woman is applying her perfume. Anti-perspirants do not work by jamming little particles into the openings of sweat pores in the armpits. That might work if sweat shot out of your body's sweat pores in miniature geysers, but on the micro-level of the skin, geysers, hoses and all the other usual ways we think of water emerging from a pore do not exist. Everything's too small. There's no way the incipient sweat water could build up a high enough pressure in its

Beautiful, but unattractive. Polarized light photo of indole, one of the many natural odor chemicals released by the human body, along with ammonia, ethyl alcohol, and hydrogen sulphide. They crumble in the air within a few minutes, but need perfume to mask them entirely. Indole itself is the smell of faeces

sub-surface tubes to flow. This is an odd limitation, but useful. If sweat gushed out of us like water out of a tube, then all parts of our body aiming downwards would be perpetually leaking – feet, armpits, fingertips, and chin – while only those always aiming up – shoulders, scalp, and little else – would be dry. Rather, sweat emerges because it's tugged out. It has a negative electric charge as it hovers in its little beads inside, and as the surface of the sweat pores has a positive charge when excited the result is that the sweat ooze is pulled out. It's like yanking out a sausage from a tight tunnel.

Enter the aluminum. Falling off from the roll-on anti-perspirant tube, landing on the crushed skin surface newly contoured by the pressure of the applicator, it short circuits the whole sweat-extruding operation. Aluminum flecks, which are the key ingredient in anti-perspirants, are negatively charged. That means the extra furry cloud of negative electrons they carry around with them counterbalances the normal positive charge on the skin surface. There's no pull on the sweat sausage any more. The aluminum is even likely to have a little left over to poke down the sweat pore tunnel and electrically repel the negatively charged water waiting deeper inside — like a hand pushing the tunnelled sausage deeper. There's a crackle, some static, the equivalent of sparks, and the whole system is shut, short-circuited and out of operation for hours. The sweat caught inside dissolves back into the body, crumbling through cracks in the sweat tubes like water from a leaky hose. (This theory is even termed the leaky hose theory by cosmetologists, because of this final water-leaking phase.)

So much for the anti-perspirant. For the deodorizing effect we expect of these concoctions a little perfume is mixed in, more in sprays than roll-ons because of the general ineffectiveness sprays have in controlling sweat due to their carrying so little aluminum. There's also a nice dose of insecticide and bactericide in the mixture, chemicals that are near-identical to the poisons in your garden shed, and which here are murder on any soft, unshelled creatures in their way. Rubbed on from the deodorant these chemicals are as acid as lemon juice. The furry and rounded bacteria normally resident in your armpits are wiped out, whole colonies coated with the poison and left to suffocate where they rest hugging the armpit hairs. Most go in 30 minutes, but as some are especially tough, it will be up to two hours after applying the mixture, say just about the dessert stage in dinner, that the hardiest and oldest residents of the environment there stop struggling and fall dead. With them out of the way there's no more odor creation, as it's their defecation of ammonia that produces the smell we're trying to avoid in using these armpit slaughtering agents.

In time though the bacteria come back. A hurried smear of anti-perspirant will have coated all the ones grasping directly on the hairs, but it will do little against the bacteria that were cunningly hanging from the hairs on slender mucus threads. These lucky ones escape in clumsy Tarzan fashion, by letting loose and falling on to the shirt or other material below. There they rest until the worst of the poison has been worn away, and then, when the bearer is kind enough to stretch or twist, will be brought up again on the shirt to start off new colonies in the armpit. Some that hid in the base of the hairs emerge too. The hastily doused wearer is neatly re-infecting himself.

Even if the anti-perspirant has not been put on in scanty quick dabs, even if it has been steamrollered back and forth in great sludgy swathes, then there still won't be a guarantee of permanent armpit peace. Indeed, such chemical slaughter comes back to haunt the perpetrator. The poison contained in commercial deodorants is so strong that when applied thickly it doesn't just kill the bacteria grasping to the hairs, doesn't just kill the ones lying on the skin surface below, but soaks into the skin like First World War chlorine gas to hunt out and get great numbers of the meekest, non-ammonia producing bacteria hiding deep down. As a result populations from elsewhere in the body migrate up and take over. These are not designed for the armpit, have no predators to keep them in check there, and so can spread to tremendous numbers, happily exuding the odoriferous ammonia as their predecessors did, and demanding even more anti-perspirant than before. Washing with soap and water avoids this problem, for it clears the region for at least a while, and always leaves enough Tarzan escapees to start over again later when the worst stresses of the day are over.

Leaving the man in such moments of self abuse, we return to the woman, perfumed and dressed for the forthcoming dinner, and with only one or two more ornamental details remaining to affix. To understand the nail polish being reached for at her cosmetic table one has to consider the dearth of billiard balls in 19th-century America. A competition was announced to provide a substitute for the increasingly rare and expensive ivory ball, and it was to win this award that an American inventor, John Wesley Hyatt, came up with the synthetic substance known as nitrocellulose resin. This proved not quite ideal for billiard balls as it explodes when struck sharply — in a later lecture Hyatt recounted receiving a letter from a Colorado saloon keeper thanking him for the new billiard balls, but noting that whenever they were hit hard they would flash, so that everyone in the saloon would pull out their gun, slowing play considerably — but with certain refinements it was good enough to found the modern plastics industry. Nitrocellulose itself survives in nail polish; watered down so that it's unlikely to explode, yet still abrasive enough to burn through the outer

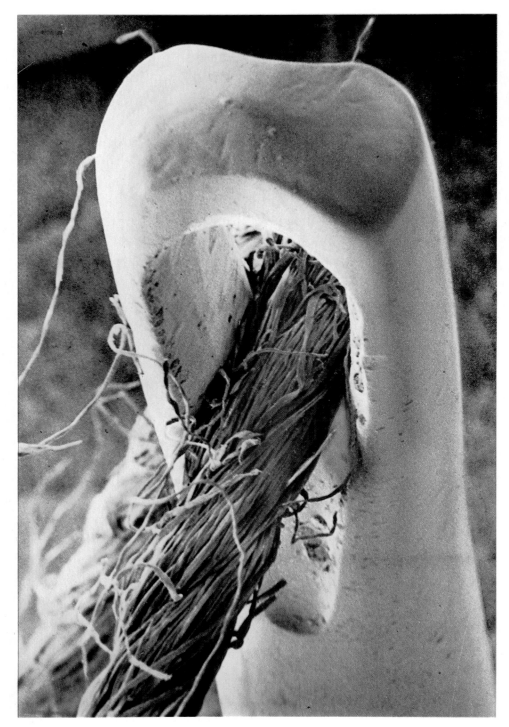

A threaded needle, containing what appears to be a bundle of ropes, but is actually fine cotton thread

Opposite: Heat photo of a man with arm raised: armpit glowing hottest (white), near tropical in heat and humidity — an ideal habitat for bacteria

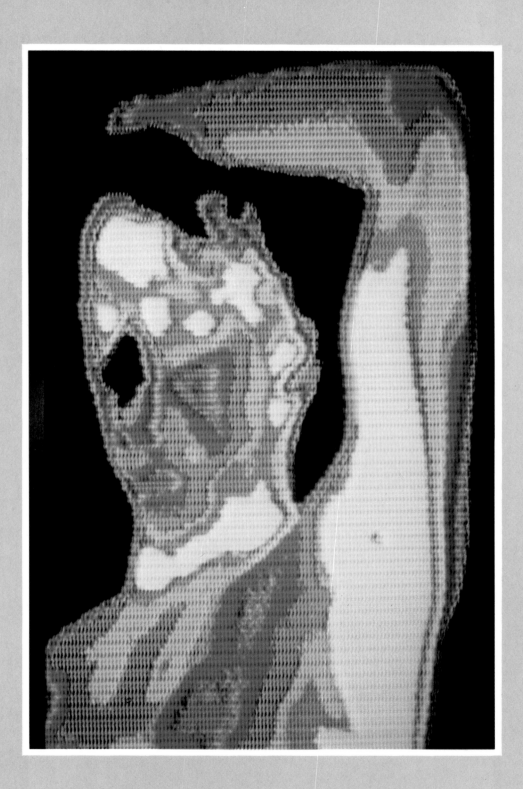

layers of a woman's nail and provide the solid covering needed in the chosen dyes.

What more could anyone want to complete the evening ensemble? A certain hint of color, worn in the ear, would do fine.

Why is a gold earring gold? A tin earring would be described as grey, not tin; a painted earring would be called blue or green or whatever color the paint was; it would not be called paint. Only gold is the odd one out, getting a color name identical to its substance name. The reason of course is that gold is one of the very few substances available to humans that never changes. A layer of 18 electrons just under its surface forms a tough barrier that won't get picked on by the corrosive clouds of oxygen we live in to form rust. What was gold once stays gold for years — centuries — to come. This has some interesting consequences. One is that much of the gold we use today was used as gold in previous ages, and only by an extraordinary mix of sales, thefts, melting down, and re-melting has it ended up in the gold earring about to be put on before the dinner party. Bits from ancient Egyptian jewellery which have been covering parts of people's bodies for several thousand years are perhaps in the woman's earring; fragments from South American mines, transported across the Atlantic in galleons to Spain, and bartered, sold and killed for in Europe are quite possibly there too. The entire world production of gold since the start of history would probably fit in a pile 50 feet square and equally high — that's all.

A second consequence of the immortality of gold is that credulous people in all ages have felt that there was something not quite natural about the metal. Objects that we see in the sky are popularly allowed to be unchanging — there's no surprise that the stars and sun and moon look the same year after year — but objects that exist on earth are not supposed to. Every substance we can touch decays over time, rusting or rotting or otherwise going off. All, that is, except gold. It is the only immortal substance we will ever touch, and accordingly should be revered. That's the theory, and Aristotle has a lot to answer for letting it slip through. The cosmology on which it is based has been discredited for 400 years — but the South African government and speculators on the Zurich gold exchange have become very rich from that continued credulity.

The particular gold hue that shines from this much re-used element is due to the fact that 'on top' of those 18 protective electrons in each gold atom there's another electron, hovering all by itself, an isolated guard, lying just beyond their protection. It's not so weakly attached that kicking a pile of gold would set it loose, but it is weakly enough attached that shining a morsel of blue light down on it will send it into the frantic, skittish oscillations one might expect from an isolated sentry startled in the night. The bump is bad for the electron, but at least in time (a few billionths of a second, it will recover and get back into place. For the blue light that stumbled across it though the collision is less easily forgotten: the light acts like a small battery pack for the electron sentry as it skitters back into position, and in that skittering the electron uses up all the energy that the light carried with it. For the light beam this is terminal. That's why if a fetishistic photographer were to steal the gold earring and put it in his darkroom where only a blue light bulb was shining, everything else in the darkroom would look blue – his hand, the pans of chemicals, the plastic cup of coffee – except the gold itself, which would look black. It uses up all the blue light that hits it because of that ridiculous electron skittering, and so has none left over to reflect.

In the more normally lit upstairs bedroom, safe from crazed photographers and the smell of developer, this blue absorption has a different result. Ordinary white light is a mix of every color, and when it falls on the gold earring every component color will reflect off – except that is for the unfortunate blue. And as any art student can testify, white minus the right degree of blue comes out as gold. That's the color of the reflected light you see: that's why gold looks, so attractively, gold.

Just a final glance in the mirror, and the woman will be ready to come down. To the beholder a reflection is often time for blank-eyed wonder, as the very origin of mirror in the Latin word *mirari* – 'to wonder at' – suggests. But to oracles, jesters and pundits through the ages a look in the mirror has been the occasion for a riddle. This is too good a tradition to let fall here, especially as our concern has been with only the latest of oracular voices, that of science. So let the beholder notice that her right and left sides seem reversed when looking in this beguiling glass. The question is: why then don't her top and bottom sides get reversed too, giving the appearance of hanging in front of the mirror upside down, golden earrings dangling up?

PART TWO

NIGHT TIME

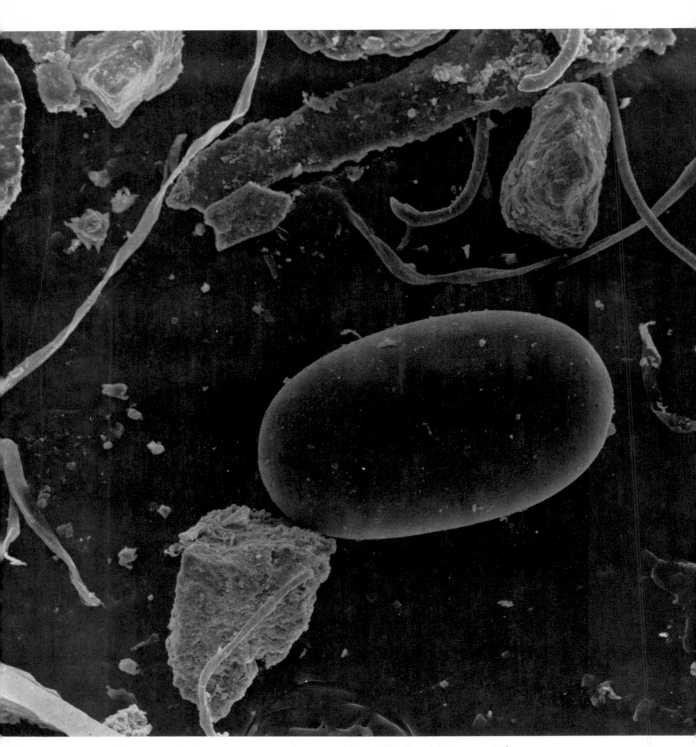

Dust: the assorted objects a vacuum cleaner picks up. The boulders are grit from brick or pavement; the cords are broken-off pieces of synthetic and natural fibers from pillows or clothes elsewhere in the house, and at the center is a cat flea egg

EARLY EVENING

FOUR

After the dressing a division of resources is declared. Someone has to do the last-minute vacuuming, while someone else will have to set the table. And you couldn't expect a someone wearing gold earrings to handle a vacuum cleaner could you?

Nobody seems to mind their own stuff on the floor, but guests can be surprisingly niggly. One educated Continental visitor paid what seemed a perfectly happy visit to a number of upper-class English friends, yet all he could find fit to say about them, commenting on the visit in a letter later, was that 'as to the floors, the lower part remains sometimes for twenty years together, and in it a collection of spittle, vomit, urine of dogs and men, beer, scraps of fish, and other filthiness not to be named'. That was in 1530, the delicate Erasmus writing, and while there has been some improvement since then there is still much to comment on. As we have seen there is a warehouse of odd items floating in the air in our homes, 300,000 or more in every cubic foot, held up by the pattering air molecules around them. These include the asbestos spikes, the micro insect limbs, the spherical chemical ash, the tire meltings, shiny cadmium blobs, sea salt, skin flakes, equatorial sand and all the rest of what we, with remarkable nonchalance, lump together as dust. The particles take hours or even weeks to descend, but with the constant nudge of gravity, fall they all eventually must, an everlasting internal rain, on our heads, tables, chairs, books, desks, lamps, stereo, clothes, shoes and, swamping all the others for being the largest surface area of all in the house, on our floors.

Devices to collect this debris have been many in invention, if few in effectiveness. The broom for centuries was best, working not as people often think by just sweeping dirt and dust forward in front of it, but also by producing a partial vacuum behind each bristle, so sucking the dust along. (Brooms wouldn't work well on the airless moon.) Controlling the vacuum, rather than relying on multitudes of moving straw bristles to do so, would be the obvious next step, but as long as everyone had the misleading notion of the broom as a dust *pushing* appliance that obvious next step was not forthcoming. The notion of improved sweeping through vacuum control would have to wait for someone who could see that pushing and sucking were just alternative aspects of one continuous circle of action. But how likely was it that someone predisposed to looking at things as one large circle would be presented with this evidence of pushing's inadequacy?

Rarely have the fates provided a better matching of man and opportunity than they did one wondrous afternoon in 1901 at the hotel of St Pancras Station, London. In attendance at an exhibition of the latest American rail car cleaning apparatus that day was one H. Cecil Booth, who just so happened to be an authority on the construction of Ferris Wheels. (The large one in Vienna's Prater Park, used in *The Third Man*, is his.) The American device on exhibit was another one of the conceputal inadequates which encumbered the era – it was a compressed air generator that proposed to clear dust by blowing air at it – but something happened when Mr Booth saw it working. From records in the file of the corporation he later founded, it appears that Booth was immediately struck with the idea that such a generator could be switched on backwards, and used to suck, not blow. The concept was so awesome that he needed to test it without delay. Booth returned to his office, knelt down on the floor, spread his lips over the carpet and proceeded to suck in furiously. He fell back gagging and choking, his mouth full of dust, ecstatic that his idea worked.

The world's first 'suction sweeper', soon renamed the 'vacuum cleaner', was on its way.

Producing a portable device on this principle was difficult. Early ones resembled small battle tanks. They couldn't fit through doorways, and were instead pulled along the street by horses, while workers, freshly reassigned from the Ferris Wheel works and wearing smart 'vacuum servicing' uniforms, walked alongside, ready to pass the long suction hose through the window into the home of any matron adventuresome enough to try this latest product of the Age of Progress. Models were quickly designed, patented, and then forgotten as new ones were designed to replace them. Finally, after a year's concentrated work, in mid 1902, Booth's labors received the finest award any Edwardian inventor could conceive. Edward VII was to be crowned that summer, in Westminster Abbey, and H. Cecil Booth, with his new vacuuming machine, was called in to clean it.

When you pull your own roaring vacuum cleaner over the floor for that quick last sweep before the guests come, some disconcerting things begin to happen. Remember first of all those gently browsing dust mites we met earlier. These were the creatures, of a diminutiveness to put them just below the limit of unaided sight, that live peacefully in your carpets and bed, feeding on whatever skin flakes happen to trickle down in the general stream of dust from above. On their size scale the carpet fibers are massive trees, and during the day the mites huddle for safety around the bases like clans of forest tribesmen. As the vacuum cleaner approaches wisps of its low pressure suction reach down from the bottom of the machine. For the mites in the path it's like a sudden whirlwind approaching out of the sky ahead. The apparently stable tree fibers are rocked and bent; the rushing air produced by the suction pulls up eye-grating dust and dried leaf-like matting fragments from this miniature world's floor. As the roar continues and the vacuum cleaner is moved back and forth directly overhead, the winds on the bottom get worse and worse. Not only are the dust pebbles being pulled up, not only are the broken matting fragments and other debris being hauled up out of sight into the sky, but even the living mite populations are beginning to be dragged away. At first it's only the stacked piles of mummified great-grandparent ancestors who go, swaying and rocking till they soar straight up in the suction wind, their hollow husks too light to resist. But then the smallest baby mites get tugged up, their eight legs grasping down as hard as they can to resist, but as the vacuum pull is great, and their weight is slight, the feet pull loose, one by one, and then the young creatures are accelerated up and out of sight in the gale too.

It sounds horrendous, but we're not mites. The baby mites whooshed into the vacuum cleaner survive their high speed ascent without harm. Their tumbling and twisting in the air is safe enough, and when they do land in the cleaner's bag their touchdown will be cushioned by the piles and piles of dust already there. Nor is this just ordinary, run-of-the-mill, sneeze-producing and forget-about-it dust. Sucked up in the house this dust has a terrific number of human skin flakes in it, and skin flakes, remember, are the favorite food of these mites. They have landed in dust mite heaven. Heads go down, and the assembled castaways feed. Continued vacuuming just continues bringing in the food. There is a slight hail from the asbestos, chrome spheres, and other incidentals in the house dust being sucked into the vacuum cleaner bag too, but those are notably smaller than the mites (5 microns in diameter to the mites' 40 or more), and anyway all mites come equipped with turtle-like armor cladding that they can pull tight for the few minutes until the vacuum cleaner is turned off and peace settles within the bag. When the captured juveniles reach puberty in there – a half week or so by our time – they will take brief pauses from their feeding to copulate, and soon have another generation of mites to join in the undisturbed nibbles. A few more weeks and there will only be a few grizzled oldsters who remember a time before the Great Uprushing; for the rest life in a vacuum cleaner bag will be the only thing they have known. (Vacuum cleaners are such an efficient way of collecting dust mites, and the bags inside such a good place to raise more, that when researchers need to replenish their lab supplies of dust mites they frequently just do some quick vacuuming at home.)

When the time comes to empty the bag in the kitchen several weeks or months later the slightest spill of dust will release some of these mites back into the open house air. Many will land on the wholly alien terrain of the kitchen floor, but for a few, tossed in slow motion by powerful house air currents in the hour it takes them to fall, their journey will take them back into the living room. There they will finish their descent, a returning lost tribe, the long forgotten exiles, floating down, down, down onto the original carpet forests their forebears started from, all those ages and ages ago.

There's another curio about what your vacuum cleaning does to the apparently quiescent floor. Because modern vacuum cleaners produce such a high-power vacuum, the air and dust they suck up is traveling at a terrific speed. (Current vacuum cleaners produce a low air pressure similar to that found at 25,000 feet or more, where an unfastened body would be sucked out of a gash in an airplane's pressurized cabin, as demonstrated in the unpleasant ending of the James Bond film *Goldfinger*.) Much of the dust is pulled in so hard that it hits against the back wall of the collecting bag with barely any slowing. That's where the problem lies. Ordinary vacuum cleaner bags are made of a waxy paper which has its fibers arranged in a grill-like network. The space between the grills is

Dust mite eating peacefully amidst contents of vacuum cleaner bag

Dust mite eating peacefully amidst contents of vacuum cleaner bag

five microns across, and while five microns is not much to us, and is even too small for the mites to fit through, it's an inviting exit for many of the other dust particles so abruptly sucked into the bag. What's pulled up through the bottom of the cleaner is accelerated into the bag and sprayed out the back again. There is even an indirect contribution from the mites who can't pass through personally: their dried faeces, which are sucked into the vacuum cleaner in enormous quantities, are shot out through these holes too. The human pushing the vacuum cleaner gets peppered with millions of high speed dust and mite pellets every minute. It's not a gentle pitter-patter either, but a wall of blasting shots, as if a fleet of miniature wooden sailing ships were blasting shells of hardened mite faeces and dust particles from their brass cannons in one broadside after another through giant nets (the bag) against him. Everything small enough from the floor is now neatly dispersed into the air; some is blasted up straight to the ceilings and starts its slow-motion rebounding from there. As an air microphysicist at the Porton Down Centre for Applied Microbiology and Research (Britain's old top-secret germ warfare center) has put it, home vacuum cleaners are one of 'the finest aerosol and dust generators known to man'.

In the other room, safe from the barrages the man vacuum cleaning has to endure, the woman's desperate rush of setting the table is underway. First it has to be wiped down, to get rid of the dust and crumbs and old jelly stains and coffee cup rings and other signs of life that have accumulated since it was last officially used. Then there's the tablecloth to spread and get even, then the plates have to be ferried across from the kitchen cupboard, then the forks, knives and spoons, and the serving platters, and the glasses, no whoops the dessert spoons are missing, they have to be laid out, *then* the glasses must be brought out, but quick, on the double, for everything must be finished before the guests arrive, it would be awful to get caught halfway, to be found out, to be revealed as not unhurried, not naturally elegant and at ease.

It is too much. The woman's fingers doing the grasping are already wet from the sponge, and with the added film of moisture from the anxiety to get everything done, they cannot do their job. As the final glass is carried, they slip. The glass wobbles and squirms, it slithers and jiggles, and then, even as the woman fights back against the pull, the glass spurts out of her struggling grasp, holds still in the air for an awful instant and then plummets down towards the unaccommodating floor.

Why it does this is a matter for puzzle. In mediaeval times they said it was because a glass's natural place is with the ground below from which it came. But that view had to be dropped when some people began to question where a drinking glass hid the memory that allowed it to recall where it had been made. And indeed if a released glass is merely returning to its origin, then it shouldn't plummet down but rather should jet away horizontally on a flight to the glass factory where it started. This would be logical, but is rarely seen.

Efforts to improve on this mediaeval neo-Aristotelian fancy have, as it turned out, been perhaps the chief driving force in theoretical physics since then, and the current view is that the empty space in your house is not the sort of smooth, uninteresting backdrop thing we often imagine it as. Rather it's a complex structure which has a very strong bend in it. The alternative word for 'bend' being 'warp', the space we're surrounded by and placed in is said to be warped. Loose objects, such as just-released glasses, track on to that warp, and follow it along wherever it leads. This seems odd, but explains a lot. A young boy you noticed in a sitting position hurtling through the sky 20 feet above you and with a box of popcorn in his arms would seem odd too, until he called down that he was riding on the new clear plastic rollercoaster and wouldn't you like to come along. Our warp is his rollercoaster.

The space warp in your dining room has existed on the plot which your house now happens to occupy for over four billion years, since the creation of the earth. It was there tripping up cavorting dinosaurs when the plot was part of a Jurassic swamp, and it was there knocking down hastily balanced hoes at finishing time when the plot was a mediaeval wheat field. Created by the presence of the enormous and relatively unchanging planet underneath us, it has had no reason to change in that time. Let those grasping fingers in the dining room tonight hold just a fraction too loosely, and the same warp will be there to lead the floating glass on down.

At the beginning there's a chance of getting it back. At the beginning, one-tenth of a second after it was plucked out of the hand, the glass is moving slower than one mile per hour. This is sluggish. With a hurried

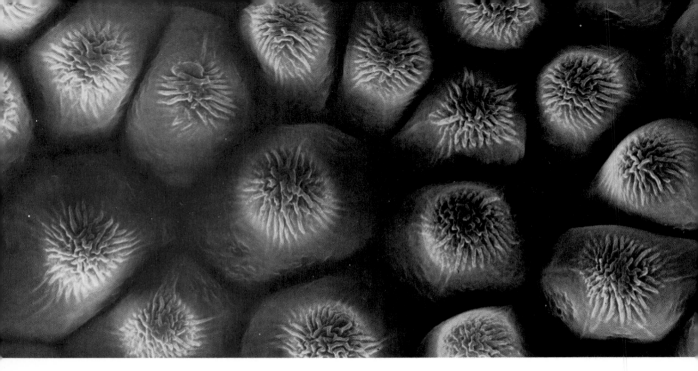

skeedaddle and lunge it could be retrieved. The longer the glass is left alone on the omnipresent warp lines however, the more it will build up speed gliding along them. That's why falling objects are so hard to catch once they get going. Wait a half second before trying to retrieve the glass, staring in dumb fascination as it tumbles floorward, and the best thing you can do is look for a dustpan. A glass dropped from waist height will, in the half second before impact, work up to a blurring 15 miles per hour, and our suddenly awakened reflexes, however wild the lunge, however clear the dire social consequences of a jagged glass heap at this moment are, can do no good against that.

Everything would be easier if the earth were made of styrofoam. The space in your house would be less tightly bent, and objects set loose in the warp would follow much gentler curved lines. To us this would appear as if they were falling more slowly, so allowing drinking glasses and even full serving platters that escaped from fingers to be snatched with derring-do just above floor level every time. If the earth were made of a much denser substance than it is now however, some kind of super-concentrated lead perhaps, then the warp lines it produced would be switched the other way, and go tighter. Then a drinking glass could only be carried to the table by lugging it with both hands, as we might manhandle a load of bricks. For the family suitcase to be manageable it would have to be packed with only

Surface of a rose petal, in extreme close-up. Like water balloons the mounds compress to give petals their distinctive softness. Puckered to conserve water, the skin contains a chemical that gives the rose its color a single toothbrush, or at most with a toothbrush and one shoe. Buttons would sag out from shirts and the thread holding them would finally snap from the awful pull; reading glasses would be an intolerable weight on the nose, requiring some sort of balloon contraption or a pulley attached to the ceiling, to keep them from guillotining through.

Lucky we are that the planet we're landed on is made of simple rock and iron, which produces a degree of space warps and tugs we can, at least on a good day, just about handle.

With the plates finally laid out, the glasses, or what's left of them, neatly in front, all that's needed to make things perfect is a colorful flower to be placed in the center. There's a quick trip to the garden, a shearing of shrubbery and a cursing at thorns, and then the woman returns, freshly cut rose in hand. In her labors she probably doesn't pause to consider the curiosity this object presents, which is a shame. Why are roses so terribly red?

The answer of course is that most roses aren't red. Wall paintings from Roman times show toga-bearing citizens enjoying the white *Rosa Alba*, in their perambulations around their villas. Even by the time of the War of the Roses there was no red rose around, which means that although the House of York did well with the true white *Rosa Alba*, the House of Lancaster had to make do with the *Rosa gallica officinalis*, which is attractive, nicely scented, and a good rallying symbol, but not red. It is pink. Modern stagings of Shakespeare, where white and red roses are hurled around, are wrong. Only in the 1860s was the red rose perfected, when the deep red but otherwise undistinguished hybrid perpetual *General Jacqueminot* was crossed with the nicely shaped and scented China rose.

Whatever the color it doesn't really matter to the bees that fertilize the plants, as they see things differently from us. In particular they can see ultraviolet color markings on a flower which are invisible to us. Looked at under an ultraviolet scope the rose on the table might reveal something like streaking white lines pointing to its center. The spectacle is what you might see looking down on an arc-lit airport at night – a bright white area with occasional runway positioning lines painted on – and that is what a travelling bee overhead takes it as too. The white glow attracts it, and the landing lights bring it in on the right path.

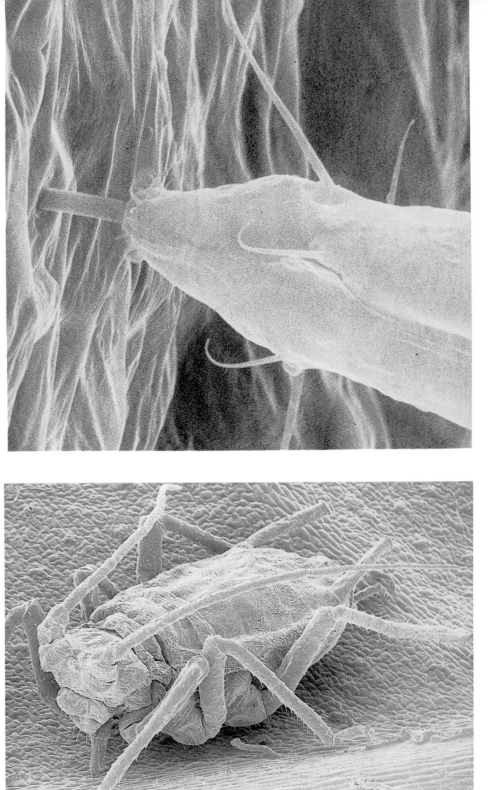

Harmony on a single leaf. Bottom left: A clambering aphid, magnified 20 times. Left: Close-up of the syringe-like stylet pushing down from its head into the leaf in search of a sugary vein. Opposite above: Close-up of a leaf — a deathfield for a wandering insect. The natural barriers include round ones producing incapacitating glue and sharp ones which impale. Bottom right: Even smaller residents on the leaf, the fields of yeast which feed on the excess sugar sprayed out by the aphids. Yeast reproduce either in pairs or individually, in the latter case popping off little buds, as visible here. Magnification 3000 times

If a rose shone entirely in the ultraviolet, as it could do, then there would be no visible light from it and we would see this floral marvel as jet black. Baudelaire might have gone for it, but the guests soon to arrive might take such a satanic item in the centerpiece amiss. The color of the rose is a fault in its design.

The graceful petals are the result of a similar design foul-up. To avoid inbreeding it's best for the rose's genetic material to be carried between different plants, but unfortunately to do so the sexual organs have to be lifted up from the plant and left dangling somewhere above it. Human sexual organs are carefully stashed in out of the way positions to avoid the dangers of daily life; a rose's sex organs face even more dangers out in the garden, where falling branches, large-jawed beetles, descending spiders, and other unpleasantnesses for the reproductive organs are around. The solution of course is to hide them, and that's what the rose's petals are for. Wrapped up tight around the vulnerables, roses are impervious to beetles, branches, or other assaults. Only for brief, carefully reconnoitred intervals will the petals fully open, and the fragile sexual parts be exposed. For those moments a strong attractant for the bees is needed, and as their flight muscles work on tanks of sugar water, an odor based on sugar will do best. Through careful crossing individuals carrying especially powerful scent oils on the surface of their petals can be selected. That is why modern roses smell sweet. If bees flew by combusting sulfur compounds, no doubt that's the natural perfume roses would waft out.

There's an interesting complication to this sweet sugar signaling by the rose, and that is the matter of the wobbly little aphids that are standing on the leaves. An extremely large proportion of garden and commercial roses have these aphids. They're hard to spot, especially the juveniles, being neatly green-camouflaged and just fractions of a millimeter across as they stand there, looking like lime-colored jelly globs with feet. Only their long elephantine snouts are a sign of a more than totally harmless nature. From this nasal tubing they extrude a narrow syringe which they pierce down into the leaf, and, once a good conducting channel is found, strain up the sugars so important in the rose's physiology which they find in there. The aphids present a pitiful sight as they feed: ungainly heads down in front, soft blubbery posteriors raised high at the back. They are quite defenseless, being only able to wobble slowly away from any attacker, or at most ineffectually kick at it. This is all right however, because the rose plant has evolved to do most of the aphid's defense for it. The thorns that are so noticeable on a rose's stem are there in large measure to keep insects bigger than the aphid from climbing up the stem and getting at these sugar grazers. And the waxy coating that you can feel on the leaves is also partially an aphid protection: malicious beetles that try to get at the aphids by landing from the air are likely to lose their footing on the wax, and slip away down the leaf.

What's going on is that the rose, with all that sugar inside it, needs the aphids to protect it. As the aphids take their micro guzzles from the rose – and this will be going on all during our dinner to come – they spray a spongy foam of water, sugar and certain proteins out from their anuses. This lands on the leaf, and while not especially appetizing for us, provides food, a foaming shower of food, for the thousands of even tinier yeast creatures that reside on the rose leaf and savor this unique meringue like nothing else. That is the key. Those yeasts hog the space that other microscopic creatures could take, creatures which would not be nearly as well-mannered as the yeasts (which after all are used to living on the rose), but would instead attack the rose to get absolutely all of the wonderful sugar it has. Such an attack would kill or at least maim the plant. So it's only the fact of the original yeasts being there, nicely fed and maintained by the aphids, that keeps those most dangerous sugar attractors away. Without the anal sprays and reproducing yeast on your rose, the dinner table would not smell nearly as sweet, nor would the rose be healthy enough to glow nearly as red.

Everything is ready, just in time to receive the guests.

Who's there? The carpet destroyers. As your guests enter the door they're covered with sand, many thousand sub-visible sand granules on average per person. It's sticking to their clothing and it's hanging from their ears, and as soon as they're inside it all begins to slip. A few sand granules slide, bump against a few more in their way, and then the cascade is on; sand falls from the shoulders, it falls from the chest: it forms an incredible haze of micro-dandruff and it keeps on falling until a great pile of the stuff is safely landed on the floor.

It's at this point that the grinding begins. The sand particles are from 1

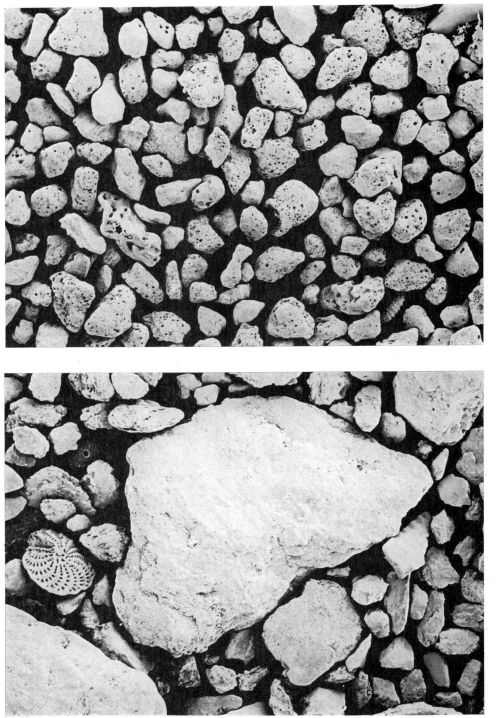

Sand. Seen close up, each grain takes on a remarkable history: some are the residue of ancient mountains, some have traveled from distant deserts, or contain fossilized marine creatures (below). The smallest particles float in the air for months, dispersing around the globe to 30 microns across, which puts them at just the right size to get wedged at the base of the carpet, right where the tufts come out. In most carpets those tufts are big chunky things, maybe a half inch (10,000 microns) in width, and against such tufted monsters the sand particles can do no direct harm. What they do is indirect harm. Carpet tufts come in those half inch chunky strands because they're built up of much smaller fibers, fibers of nylon or wool or whatever the carpet is made of, and those fibers are not 10,000 microns across, not even 1,000 or even 250 microns, but rather ultra-slender delicate things of only 4 or 5 microns across.

These are what get the sand. It falls next to the fibers, and in other circumstances it would rest there quite peacefully, doing no harm, but at this moment, with the guests who just brought it still standing there above, nervous and anxious and keen to prove their worth, then the sand is made to do something far less nice. Anxious guests are burbling, gesticulating, gushing guests and individuals in such social *extremis* never, ever stand still when they can twitch. Not wild twitchings, not *grand mal* seizures or Ray Charles stomping, but little gestures, weight readjustments, knee stretches, slow-motion two-steps, and, cruellest of all to the carpet, full-weight, single-footed shuffles.

A shuffling guest will lean on one foot, jam it into the carpet, and then, without even thinking, without guilt, without contemplating an offer for carpet replacement costs, will rotate his entire body weight through several radians of a circle on top of that pivot. Think of what that does to the fibers underneath. They're squeezed onto the sand particles, they're forced and pressed and jammed, and as long as this awful rotation carries on above them, they get it. The sand hacks the fibers, it gouges-scrapes-mutilates the fibers, and when the gigantic weight overhead finally removes itself, finally decides enough twitching has been done on that side and it's time for a little nervous shuffling with the other foot, there is no longer a landscape of smooth, fresh clean carpet fibers there but a wreckage of battered, pulpy, half slit-open things. With each twist the horror is repeated, and while one sand particle grating away on the fibers in its range is not going to do much overall damage, a cascade of thousands will add up.

Anxiety carpet destruction is a well-known phenomenon among floor covering specialists. In addition to front halls, scenes of such social unease, the area in front of office reception desks is especially liable to this attack, as that's the other place where people go through their worst feetly contortions, cringing, anxious, not knowing what's to come. The specialist literature includes suggestions to vary the placement of the reception desk, or even employ friendlier, less fear-producing receptionists, so as to keep down the grinding, and avert the ultimate sign that a floor covering consultant has failed at his task: the installation of a square of clear plastic sheeting in front of the desk.

Sand of course is only one constituent of the incredibly varied dust that guests carry in on their bodies and on their shoes. Any bits of grit will do when it comes to destroying carpet, which is why it's so important for the continued life of a carpet to vacuum them away thoroughly and regularly. But we will concentrate on the little-known granules of sand here, letting their story serve as an indication of the convoluted paths all the other floor-attacking particles have taken on their way to the front hall at greeting time.

The first question is how did this sand get on the guests in the first place. Was it a hasty trip to the beach, a crawl Dune worm style through the dunes and then, scant attention having been paid to the hour, a frantic drive along to the dinner party, without time to go home, brush off, and change? It was not. All your guests come with this sand, not just the inattentive, beachcombing ones. It landed on them from the air, coating them during the day, or just on the way from car to house, and it got there, curiously enough, from take-off points in the distant deserts: the Sahara, the Gobi, the Mojave, and other such miserable uncarpeted places. Our fiber destruction is their fault. This is not a question of giant sandstorms, howling sky-clouding things, as those are too rare to have much effect. Rather it's a matter of the frequent light breezes, 5-10 miles per hour. While those are great for sweating nomads, desert rabbits, and other locals, they are what start off the train of events that lead to our later carpet misery.

The reason is that desert sand is not simple sand. It comes in different sizes. When a breeze starts it doesn't lift the 1 mm-broad giant sand particles, or the 0.001 mm tiny ones, but rather it works on the intermediate sized particles, the ones $\frac{1}{30}$ mm or so across. In the Sahara there are lots of those. The breeze knocks them forward, and they begin to roll. As the breeze continues they continue to roll, tumbling like miniature sage brush across the ground. Unfortunately for our carpets, they don't get to just continue rolling smoothly across the surface till they're out of sight. If they hit one of the larger sand particles projecting out of the surface they will roll right up it and take off into the air. The effect can be imagined as what would happen to a rolling sage brush that came across a water-ski ramp jutting up in its way, left over from an abandoned movie set perhaps. The sage brush would roll up the ramp and go flying off; as there are almost as many larger sand-particle take-off ramps in the Sahara as there are medium particles to roll into them, plenty of the tumbling sand particles are sent skyward this way too.

At this point the analogy with the sages brushes of Western extra fame must be dropped. Sage brush is hollow, and sand particles are solid. Think rather of large concrete spheres hurtling up the ski ramps and taking off. The concrete boulder would tumble as it soared, and when it landed it would knock the stuffing out of whatever was below. The medium sand particles do the same, spinning as much as 1,000 times per second while they're airborne, and then plummeting down. Landing on big sand particles they just crack open, but crunching on the smallest 0.001 mm sand pieces, they send those rebounding up skyward as sandy dust.

It is difficult to emphasize just how many of these tiny sand pieces are released each day. It adds up to about 100 million tons per year — fifty pounds of the stuff strewn airwards for every person alive. They rebound in an invisible haze several yards above the deserts, and are so small that any whiff of rising air will send them even higher. And there is much rising hot air in the desert. An immensely large number of the smallest sand particles are forced up into the layer 3 to 5 miles above the earth's surface, and from there many will drift away perhaps to cross half the globe. (Raging sandstorms will send the stuff higher, for dispersal in greater quantities, but this, as we mentioned, though spectacular when it happens, is relatively rare.) Almost certainly the last airliner you were in met clouds of millions of these tiny sand pieces. They were too small to see out the window, but the workers who have to repaint the airplane frequently, to a great extent because of the pitting they produce, can testify to their existence.

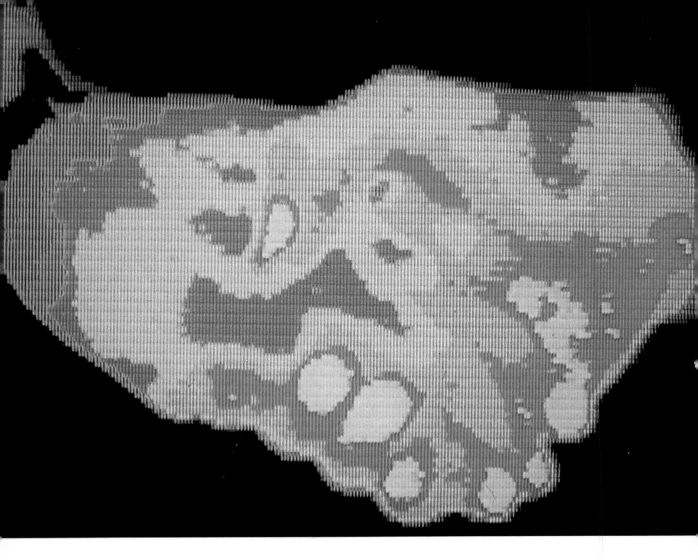

After several weeks the floating desert sand sub-particles start filtering back down, pelleting your unexpected guests during the day or even just on their few steps from car to house, and eventually being dropped off their clothes and body onto the final resting spot in your carpet. It's unlikely that in the middle of greeting the guests you would throw yourself bodily on the carpet and examine it under a microscope for the fragments of Gobi, Sahara, or Mojave embedded within it. But if you did, there they would be.

What happens next is rarely open to choice. Even while the sand grinding is underway the guests will likely begin that intricate forelimb clench and joint sinuous movement which we call the 'handshake'; then after that there will be an interval for the guests to recite items from the

The topological contortion of a human handshake, seen in infra-red. White is the hottest; then red and yellow; black is coldest. The pressure of this ritual grasp squeezes finger nails to make them glow as hottest white

EARLY EVENING

standard litany of greeting phonemes, and the hosts, after more due pause, to contort their faces and let loose from the buccal cavity that explosive low frequency gasping sound we call laughter in response to the aforementioned litany.

It is a soothing ritual, and accordingly near universal. But it only works if everyone is willing to take part. If one of the guests in the hallway is a dazed-out guest, a guest who should never have dragged himself there, who should have stayed at home, rested in bed, infused medicaments and high strength chicken soup; if one of the guests is a guest who is sniffly, rheumy-eyed and suffering from a cold, then the greeting ceremony will likely be cut short by that most violent of all inadvertent anti-social acts: what certain Northern European peasants in the era of the Bubonic Plague were horrified to call the *pfnusen*, and which we, updating their word but keeping their distaste, call the 'sneeze'.

Having a cold means that you have lots of extremely small virus 'creatures' living in your throat and nose. (Viruses are even smaller than bacteria, containing little more than nucleic acid surrounded by protein, but capable of reproducing and propagating themselves when landed on a suitable target cell.) These viruses lead to an urge to sneeze, and however well brought up you are, however much you realize the social opprobrium such an explosion in confined quarters will produce, the urge to let loose soon becomes well nigh irresistible. The microbiological evidence is that viruses have evolved irritant proteins on their surface so they can trigger off a sneeze that will propel them to the fresh feeding grounds of new human hosts. A wadded-up old kleenex may be frantically jammed against the nose, but for a sneeze that wants to get out that is no defense.

Our guest's sneeze will speed out of his nose at 40 miles per hour or more – that's a Force 8 gale on the Beaufort scale. This compares with the gentle five miles per hour – Beaufort Force 2 – of normal breathing. Now while a kleenex is perfectly designed for feeling soft and reassuring when dabbed against a sniff, it's not at all well constructed to withstand the dual nostril blast of a sneeze. Beaufort Force 8 is, as the shipping manuals attest, capable of 'breaking twigs off trees'. There's no need to mention what it can do to the guest's mere tissue. Just to make matters worse, a well-used kleenex is going to be a wet kleenex, and wood-pulp based products (as the kleenex) when wet drop to only 30 per cent of their usual strength. (It would be even worse, but for the stiffening resin that's added to every one of the estimated 300 billion plus paper tissues sold yearly.) And then there is the problem of the holes.

A paper tissue is more hole than tissue. Indeed a kleenex is soft precisely because its fibers are so far apart; woven closer together the kleenex would be a more sanitary object, but unfortunately feeling like a crinkly plastic sheet it would be a less saleable one too. The smallest nasal spray globules speed right through those market-enhancing holes. In the spray globules are the living virus creatures. One second after a guest's sneeze they will be through the tissue and already 12 to 15 inches in front of his nose, and in that time the spray globules they're riding on will undergo a curious change. Blasted in a 40 mile per hour wind they quickly dry out from the friction as they gush through the air. All that's left of the soft water globules is a flotilla of dart-shaped mucus residues. On these hardened carriers the viruses travel far, gliding horizontally across the front hallway, quite as alive as they were in the nose. They're going much slower now than at sneeze release, and have at least a 30 minute flight in store before they hit the far wall. If, that is, there's nothing in their way. Let two horrified human dinner hosts be standing in line with the flight path, and the darts will smack into them instead of the wall. They will land on the shirt, they will land on the face; they will crash-land on whatever part of the hosts' bodies they can reach.

Is there a defense? Since the virus-laden darts are sprayed out straight through the kleenex, standing somewhere out of direct line might seem right. Unfortunately, the viruses that enter your house in a guest's throat and tickled nose are not so easily foiled. While much of the sneeze blast does go straight through the hastily raised kleenex, a certain amount of it is blocked, and spurts out through the sides instead. These turn into gliding darts too, and disperse through the entire volume of the room, a rain of aerial cold viruses, a massed bomber attack that seeks out and finds the horrified hosts, even if they've stood back, even if they've edged away, scampered sideways, and hoped that these precautions would spare them from the so briefly pre-signaled spread. Leaping directly under the sneezing guest's legs and cowering there till the worst was over would provide some safety, but could provoke gossip. Better by far is to bundle the whole party out of the front hall, where the gliding danger everywhere lurks. In an hour or two the room could be safely re-entered, all the virus-carrying gliding darts grounded on the floor, where they could then be happily crackled underfoot. Until then generous offers of fresh kleenex, and as wide a berth of the virus-ridden guest as etiquette permits, will have to do.

For humans lucky enough to live in heavily populated countries, virus spread on salivary darts is not all bad. Everyone gets peppered with the stuff, and indirectly that leads to almost everyone being fairly immune to most of them. For individuals from more sparsely populated regions it's not so good. The conquistadores who landed in South America in the early 1500s and proceeded to destroy great civilizations that had been extant for centuries did so not because of some great superiority in Spanish weapons of the time, but merely because with each sneeze they released great numbers of contaminated gliding micro-darts that the Aztec and Mayan locals had no defense against. Had the situation been reversed, with the South Americans having lived in more crowded sneeze-ridden urban quarters than had the Europeans, then the conquistadores would not merit entry in our history books, or if so would appear only as defeastadores, whose ships and foolishly not-hidden maps were what allowed the Aztec invasion and conquest of Western Europe to take place.

It is these virus-emitting and carpet-crunching beings who will be guided from the front hall into the living room for the necessary decompression stage before dinner proper can be served. Only with the guests settled down there can we finally get a good look at them. It is a disturbing sight, just a vista of dead cells, that being what their face, hands and hair are covered with, but it is made a little bit better by the fact that what we see is at one remove from what the guests actually are. Human bodies do not emit light, nor do stockings, shirts, blouses, scarves, and other accoutrements. There is nothing inherently bright or colorful about humans, as the experience of sitting with others in a closed room with the lights off will demonstrate. The entire visual image you get from the two guests, now sitting and perhaps even still grinding in their chairs is due to the way they reflect ultra high speed light photons that are emerging from the switched on tablelamp.

This is such an odd notion that it's only fair many early scientists didn't bother to consider it. Light comes from the human eye, they sensibly

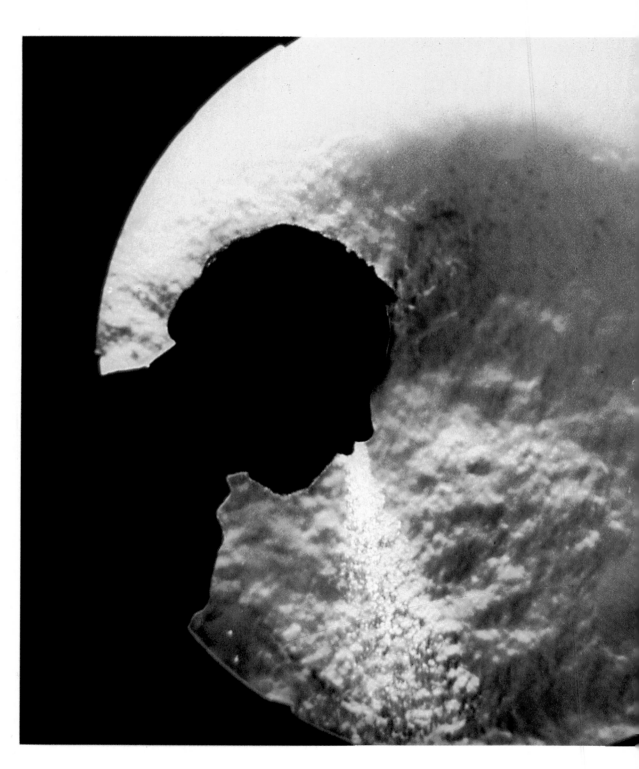

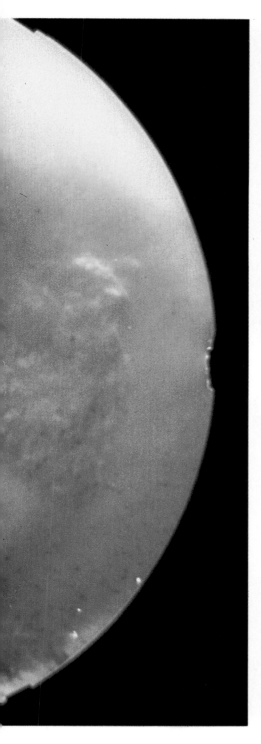

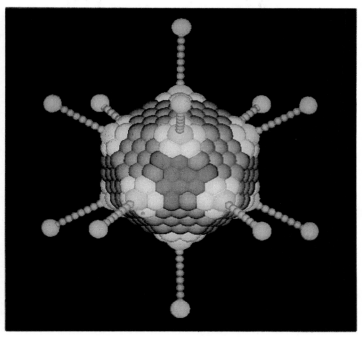

Why microbiologists get scared to go out in public.
Above: A cold virus, in computer image. Left: A sneeze
comes full out, and (below) attempts at restraint

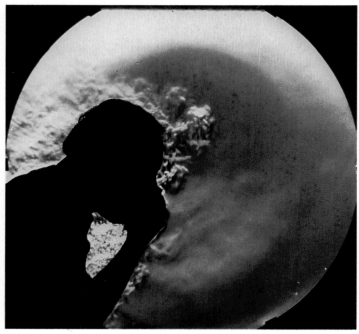

reasoned, and things got bright because the eye was pointed at them. Drawings of the time show draftsmen and laborers going around the streets of Renaissance Florence with Superman-style light cones beaming out of their eyes. It was a plausible idea, especially as the drawings always showed street scenes during the day and, as everyone knows it's impossible to see people during the day without pointing your eyes at them, these invisible light beams from the eye no doubt swinging along like searchlights as the head was moved. Unfortunately it's also impossible to see people on an unlit night, even if you do swing your eyes and suspected Superman-style beams along with you. The theory was dropped, and the current one, though the details would take a few centuries of working out, took its place.

The illumination of your otherwise invisible guests in the evening living room is especially notable, as the light bulb doing the illuminating is just a few square inches in surface area, and the guests, even after a strict diet and situps, have a surface area of over eleven square feet each. Do all the necessary photons come from that one little bulb? It would be a problem if photons were like little balls, each trundling across the room from the light bulb and splatting the guests with a volume of bright light to reflect. Then the bulb would have to be packed pretty tight to hold all the necessary photons inside. The answer to this photon packing problem is that the photons which hustle out of your light bulb are not little balls, are not even little-bitty balls, but are quite some other object: a particle weighing nothing, with no mass at all, but happily capable of carrying a certain amount of energy with it as it trundles through the air. That energy striking the guest's face and clothing is what rebounds and lands back at your eyes. Or rather, some of it rebounds and lands at your eyes: most scatters wastefully in all directions, spattering against the walls, the carpet, your shoes, your nose and every other surface imaginable. At each of those points a full image of the guests would be visible – there are enough photons flying around for that – if anyone were there who cared to look.

This extraordinary waste of wrongly directed photons is only possible because the filament in your light bulb is continually supplied with new shipments of them to pour away into the room. They come from the power station that's at the far end of the house's electricity supply system, but since they're such tiny particles, since they weigh nothing at all, there's no need for them to be jammed into the circuitry at the power station and physically pushed along, like so many message containers being forced along pneumatic tubes. The circuitry in your house walls is not full of photons on store. All that comes from the power station is an electrical field strong enough to make photons plop out and then fly away from the atoms that make up the filament inside your light bulb. As long as the atoms are steadily resupplied by energy from the power station there will be no problem for them to keep on sending out as many fresh photons as the viewing humans in the room could wish. It's as if a peaceful chicken were cajoled to stand between two electrodes until, feathers frazzled and eyes askew, it came out with an appropriately luminescent egg. More cajoling, more luminescent eggs: there's plenty of light for everyone.

There's an intriguing complication. If you wanted to tell from exactly what orifice a startled chicken laid loose a luminescent egg, all you would have to do is bend down and look, or, to make the analogy more exact, since we assume it was dark before the light was switched on, all you would have to do is shine a flashlight at its bottom to light up the scene, and then wait to see where the egg emerged. This wouldn't work for a filament atom in a light bulb however, because the glare of a flashlight would make it impossible to see where an individual photon was coming out. Shining only a few photons would still produce glare, and if you go all the way and shine just one photon then you're in another mess. Any photon you shine down is going to be just as big as the photon coming out. (You can't shine half a photon: they only come in wholes.) When your illuminating photon hits it will knock the plopped-loose photon out of place, and once that happens you've lost your chance of seeing where it came from! Shining no light wouldn't help, because then you would literally be in the dark.

It seems to be a dead end – either you try to see the photon and disturb it, or you don't disturb it but then don't know where it is – and the only way physicists have been able to preserve their sanity is to say that there really is no specific place on the filament atom where the photon came out from. This is the famous uncertainty principle. It's as if the best an anatomist could do were to tell you that a chicken lays eggs either from its mouth, or its posterior, but beyond that he couldn't hope to be more

specific. The photons that come out from your living room light bulb and end up on your guests each come from the surface of a filament atom, but what exact part of the atom they came from, or indeed if there is an exact spot they came from, is a total mystery to science. Although the man who first discovered this result, Werner von Heisenberg, was a pretty unsavory character – he ended up heading the Nazi attempt to build an atomic bomb for Hitler and the master race – the argument he used has been unable to be faulted, and after some initial resistance is now accepted by all scientists as a basic part of nuclear physics. The only notable objector, who continued holding out against the result years after everyone else had accepted it, and was forced to live in intellectual isolation because of that, was Albert Einstein – a resistance and isolation especially poignant as it was Einstein's own discoveries on photon emission in the early part of the century that made it possible for the young Heisenberg to come up with the uncertainty principle.

Sitting nicely the talk will soon begin, the rush of words, the torrent

Art Deco in the circuitry: thin section of electric cord insulation. Copper or aluminum conductor goes through the center; plastic or rubber wrapping on the outside is impervious to electrons in the conductor, and so keeps the current from dribbling out and burble. Ten distinct speech sounds are released per second in normal conversation, and the rate will reach 15 or even a bit more in those especially impassioned moments when a conversation partner's comments must be interrupted, corrected and put on the right track before he can be allowed to continue again. Consonants are the worst in all this rush of speech, for they're short little sounds, quiet things, with hardly any acoustic power. Vowels are much easier to make out, being great hulks of sound, long single frequency cries. In refined speech there will be lots of consonants, wafting in pitiful weak pulses across the room, but when there's need for something more than refined speech, when the speaker has been delicately emphasizing a point and spills a glass of wine on his lap in the move for example, then the delicacy of the consonant will give way to the power of the vowel: cries of anguish and anger are invariably based on vowels, kept open at the same pitch for long, expressive intervals.

The reason for the difference is that vowels are easy to construct. They are created by keeping the breath channel unblocked, as indeed the etymology reveals. 'Vowel' is simply an extension of the Latin *vox*, the same root as for our 'voice'. People who have nothing to say can easily give the appearance of being knowledgeable, or at least can dominate the conversation, by simply refusing to close their mouth after a vowel has been started. The vowel sound, being so simply generated, will just keep on coming out, whence the famous braying of English county hostesses, or the interminably held greeting ejaculations of Hollywood rivals. More usefully, an over-emphasis on vowels will make a complex sound easier to follow. This is the reason nicknames are not just short names, but are vowel-enriched short names: Richard turning into not Rchrd, but Dick; William transforming into not Wllm but Bill. It's also why bestowers of diminutives are so keen to ignore final consonants and replace them with a vowel, or, if that dropping isn't possible, to tack on a nice, easily comprehended vowel instead, whence all those once distinguished Davids becoming Davys, the Rons becoming Ronnies, the Dicks becoming Dickys, etc.

This horror of the consonant, this urge to excise them at every opportunity, comes because they're so much harder to construct than vowels. Complex throat configurations, often with contortionate tongue

The filament inside a light bulb. An electric field stretching back to the power station curves around these wires and makes them heat up, so that they glow with visible light, like a miniature branding-iron

help, are needed to make a passable consonant. Being so painfully constructed, consonants are also feeble and hard to hear. However. It is our unfortunate lot to have to depend on these consonant sounds to get the meaning of our words across. Non-tonal languages such as our European ones are not clever enough to get full use out of their vowels, to modulate them up and down like various oriental languages do, and so are stuck with having only five or so different basic vowels in their repertoire, the famous a, e, i, o, u and sometimes y. Now five and a half vowels do not a distinguished language make. Only by hooking those basic vowels up with lots of other more complex sounds can a suitably interesting vocabulary be made, and those more complex sounds, the only ones humans can produce short of carrying hollow gourds and zithers around, are none other than the breath channel distorting, quiet, and hard to make out, miserable consonants we started with. Only with consonants does meaning come clear, as the exercise of deleting all vowels from the first phrase of this sentence will show: only 'wth cnsnnts ds mnng cm clr'. The same phrase with consonants dropped would be only 'i ooa oe eai oe ea!': an utterance which, however heart-felt it might be cried, however violent the arm gestures used to accompany it, is lost in the depths of incomprehensibility.

These are the miserable objects that must be received if you are to make any sense from the gush of sounds a dinner party still in its living room stage will produce. Whether we hear them depends on how reverberant our walls are. If our walls are soft porous things, if the consonant-bearing sound waves that slosh onto them get caught in micro niches in plaster or curtain and use up their energy heating the air caught there, then there won't be much stray sound reflecting back off the walls. It's not that there will be zero reflection off softer surfaces — for that you would need what the acousticians quite seriously call a surface of 100 per cent OWUs (Open Window Units), or no walls at all. Rather if there are enough pores in the walls the sound will be so much diminished as it travels from surface to surface that after just a few journeys it will be so weak, so much power having been chewed out of it by that fruitless air heating, that it will be too quiet to be audible. After that point its continued existence might be of interest to the metaphysician, intrigued by the fact that every phrase and laughter released into your room lingers

Left: Not an aerial shot of a highway going through the Utah desert, but the interval between two series of tracks on an LP. The rolling sagebrush are dust particles. Centre: The tone-arm tugging along the grooves with (below) the diamond stylus fitting snugly in place. A straight groove produces quiet music: a wavy groove is louder

on in dimmer and dimmer reflections until they're finally lost in the background random heat vibrations of the air. But for all other purposes it can be ignored. So long as the reflected sound reaches inaudibility within $\frac{1}{20}$ of a second of being produced – what works out to it taking no more than 36 feet of rebounding flight around your room – then it will not seem to distort the direct sound, with its crucial consonant message inside. If the walls did not chew up enough sound, if they sent back loud and clear rebounds even after this $\frac{1}{20}$ second interval, then what you started out hearing as a meaningful consonant would get merged with the rebound signals and turn into a meaningless blur. We'd be back to having to decipher the guest's gasped 'i ooa oe eai oe ea!', and however well intentioned you might be, such communications do make prolonged conversation difficult. Open plan offices are so loud because they get filled with leftover rebounding sounds – nail files rasping, hushed talking, paper reams shuffling – all ricocheting in limbo above the cubical partitions and typewriters, all arriving later than that crucial $\frac{1}{20}$ second and so giving the impression of arriving in a constant aural blur.

The history of music can be seen as the history of making yourself heard in rooms with a worse design than open plan offices. Mediaeval cathedrals were made of stone, and since that material has hardly any of the air-containing absorbent pores needed to soak up sloshing sound waves, the only kind of music you could have in those structures were simple chant intonings, where one voice held without rhythm over an entire line. To us this is monotonous, but in structures where reverberation times could be up to 10 seconds, anything more snappy would have been impossible to hear straight. As long as music stayed in these great stone churches it continued to be limited by the reverberation times. That's why Palestrina's choral works, from the late 16th century, have little sense of meter, and are lines of sustained overlapping notes – they had to be, as they were still being composed for the old churches. It was only in the following century that a new possibility arose, and that was because of the development of what were the most avant-garde, high-tech structures of the day, the Baroque churches. These were often smaller than the old structures, and also had gilt trimming, sheets of white paint, and lots of twirling wood designs. All were loaded with the micro pore structures that soaked up otherwise reverberating sound waves, and so made it possible for young new composers, and eventually Bach, to work out new types of faster, polyphonic music.

In Italy there was something even better coming. The first large opera houses there, building crammed with sound-absorbing plush seats, private boxes, and ever more wood, opened the possibility to even faster music. Mozart and his immediate predecessors did wonderful things with it. They started composing for lots of flutes and violins, tiny instruments compared with the fist-thumping organs popular before, but instruments which precisely because they were so small could be used to produce lots of short, fast notes. With a time set to 132 crotchet beats per minute, for example, a flute or violin can easily produce up to four notes in each beat; an old-style double bass player would be working very hard to send out even one note per beat at that rate. To some ears this was unfortunate – whence Joseph II's famous remark, 'Too many notes, my good Mozart, too many notes' – but to many listeners, especially comparing it with the simple contrapuntal stuff that had come before, it was good enough to commission more. (Catastrophe comes when you tried to play the new fast stuff in the old stone churches, which is why Prince Charles and Lady Diana Spencer had only carefully slowed music at their wedding in the reverberant St Paul's Cathedral.)

Yet this setting of photon and sound wave is only sampled, the favorite anecdotes are only begun, before the announcement comes that all has to be stopped, anecdotes abeyed, while operations are transferred to yet another chamber. For dinner, at last, is to be served.

In the dining room, a transformation occurs. Pretence is dropped, and gluttony revealed. The guests grab at the hors d'oeuvres, keen to swipe the nouvelle cuisine suggested crackers covered in peanut butter, the celery sticks and the best of everything else on the tray. They struggle to chew fast, to swallow it all, for there are more starters on the table, and unless they get it all down it might be removed, pulled from their grasp and carted back into the kitchen before they have even had a chance! Desperate hands clutch at bread rolls, piggy eyes search for the butter to spread on, and if by a horrible mischance the wine starts to be poured while their mouths are still full, grunts are bellowed out, earnest mouth-full grunts, to ensure that their glasses, their receptacles, do not somehow get missed in the general filling up. Gusts of moisture rise from the sweaty faces (several grams of water from two strenuously eating individuals in 15 minutes); chewing, slurping, and slobbering sounds fill the once-gentle air. The rheumy-eyed infirmity of having a cold is left behind.

It's an unfortunate spectacle, but it was worse before. Etiquette manuals reveal the sad truth. Mediaeval kings, chivalrous knights, and the fine ladies who were the objects of their love; all had to be repeatedly told not to spit on the table, not to clean their teeth with a knife, and not, once the knife was laid down, to attempt to continue the tooth-brightening operation with the tablecloth. It was further clearly noted that the habit at court of grabbing the food with both hands at once, though understandable, was also to be suppressed. The correct procedure was to tear at the meat with three fingers only, and if too much was crammed in the mouth at one go the excess was to be discreetly spat out on the floor, not onto the table or, as was apparently a frequent event, back into the serving tray.

Such behavior continued well into the early Renaissance, and was due, in large part, to the lack of that three-pronged culinary implement we take for granted: the fork. Forks were unknown for eating at table even in Italy until the mid 1500s (Da Vinci's *Last Supper* is drawn without them), and only spread to backward northern lands such as Britain and Prussia at the end of the 1600s. There were problems among the English upper classes in the transitional period: sometimes the food would be lifted by hand, in the old, comfortable fashion, and only when safely grasped would it be stabbed with the fork for the final journey to the mouth. Indeed the whole concept of the fork represents a mechanically-induced distancing of the body from the outer world, and only became widely introduced at the same time as those other famous artificial dividers of the body from the outside world, the handkerchief and the pyjamas. But about those, more later.

Even the table which is the focus of all these ministrations is a surprisingly recent invention. The reason is that tables are heavy, and in earlier times people, even the great lords, were on the move so regularly that they couldn't bear to cart such heavy objects around with them. They had to stay on the move because they were poor, on the level of the better Third World peasants today. One solution was the individual folding table,

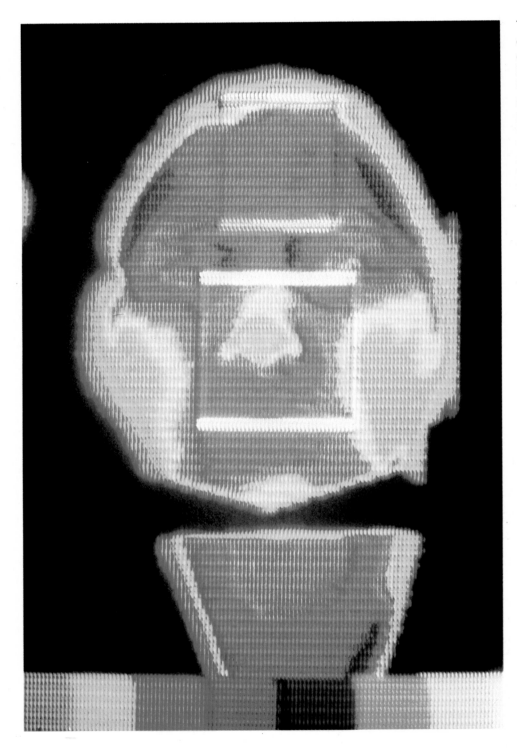

Take an after-dinner drink of alcohol and this is what happens. Left: Thermogram of a person's face before drinking. Color strip shows white warmest down to green and blue coolest. Opposite: The same face just a few seconds after downing a drink. Now it glows with heat — the cheeks rising five degrees

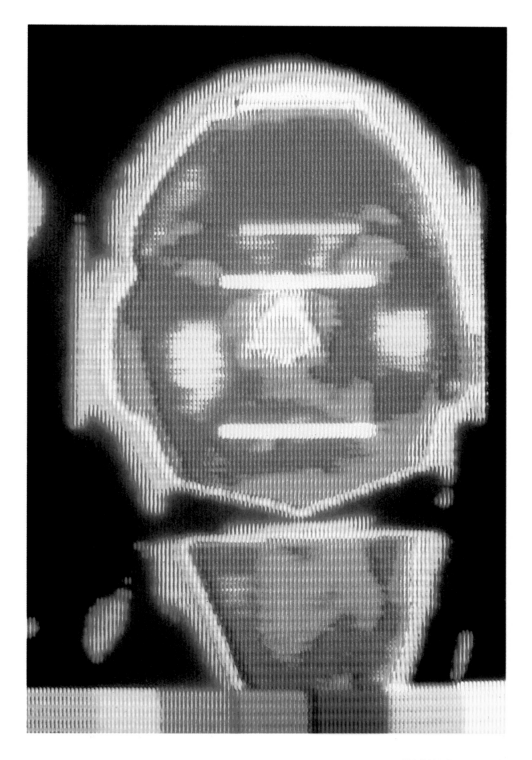

something like the one that reappeared in American suburban homes c. 1960 to hold TV dinners. These much-abused devices have a noble pedigree, as the contemporary records show that French and English nobles would almost always tuck their hosed legs under such individual TV tables when it was time to eat in their *châteaux*. If they did have to set up for a full sit-down bash, with lots of guests coming, then they would just clear away these tray tables and, once the guests had come, set up a flimsy arrangement of planks on a trestle base. There was no way to do it before the guests arrived, because few nobles had the wealth to have extra trestles and planks around. Guests who wanted to eat had no choice but to schlep their table with them.

Around that impromptu construction everyone would sit in little folding chairs, something like the contemporary film director's chair, only without such comfortable canvas in the fold. Here again the reason was portability: folding chairs are easier to lug than solid ones. The only place solid furniture existed was in the church, and even there it was likely to just be the odd oak table or two, where Communion could be performed. Even the great cathedrals were too poor to supply chairs for everyone, whence all the kneeling and bowing in traditional rites. Our mistaken image of mediaeval feasts being held around great tables probably comes from Hollywood films, a shame actually, for if the director had just turned the camera on himself we would have seen the sort of furniture an economy of scarcity and constant moving produced: the director's folding chair, his lackey's bench stools, and the easily carried folding table in the corner for the lunch break.

Setting up the table was a special art. The trick was to see that all the eaters were lined up in a row on one side of the table. This way their backs could be to the wall, a precaution against mugging, garroting and other common mayhem. Such precaution was necessary. With all the visitors coming and going, plus all the servants the visitors brought, and the families that the servants brought, it was hard to ensure that no undesirables were in the house. In the Black Book of Edward IV strict orders are given that the King's bedding *must* be locked up promptly each morning, as he was getting fed up with it regularly being stolen. And this was in the greatest castle in Britain. Our only holdover of this seating arrangement is probably the one-sided row of politicians on a raised dais at fund-raising dinners; the arrangement we use at less exalted occasions, with people sitting across from each other and not everyone guaranteed a back to the wall, descended from the rough all-in piling of the servants' wing.

In the glutting at the contemporary table little attention is usually paid to what's actually within the delectable morsels being shoveled up. The peanut butter on the crackers is gobbled without delay, the chewer oblivious to the average of two 'insect parts', most often the torn-off legs of grasshoppers or field spiders that were unable to hop or crawl away fast enough when the harvesting equipment came through, which government regulations accept as likely to be contained within. Nor is there worry about the living fungus bodies, great writhing colonies of the stuff, safely nestling within the cavities of the Roquefort cheese, let alone the huge numbers of bacteria, swimming, gliding, bouncing, and plodding through everything else. Only a microscopic examination could reveal those, and a good guest would, unlike Pasteur, not think of embarrassing his host by bringing such an instrument. It could also detract from valuable eating time. Rather what a properly sensitive mouth-stuffing gourmet will do is lean over and try to get a sniff of what's presented before him. This is called osmogenic sampling, and seems to be a natural impulse of the human body when presented with food-like odors. A well-trained nose can detect much. Are the hosts trying to foist off any of their old scrapings of butter, leaving the best and fresher stuff for themselves? This is easy to tell with the nose test, for butter when it starts to go rancid produces certain split-apart fatty acids, which as it turns out are identical to the highly odorous sex chemicals a female dog produces when she is in heat. Stew too is worth a good sniff – considering that it originated as a way of softening old pieces of meat that otherwise would be too far gone to handle – and while one's at it the bread, the soup, the mashed potatoes, and just about everything else are worth taking a quick sniff of too.

Wherein lies the problem. It is very difficult to take a quick and unobtrusive sniff of the food that's being offered you. Few of us would be so uncouth as the guest described above, but still whenever we're put in a setting where there are a lot of new smells, especially food smells, then the urge to sniff them, to test them, to get even just a little whiff of them, is well nigh irresistible. And such an osmogenic sampling urge is something that's hard to carry out without making a spectacle of oneself. Odors are detected only deep within the nose, and to draw the air containing the odors there we have to pull hard. Normal breathing sucks air into the nose at a lowly four miles per hour. But the wind speed needed for a good sensory sniff is 20 miles per hour. This is Force 5 on the Beaufort scale, enough, as the mariner's definition puts it, 'to raise dust and loose paper, and make crested wavelets form on inland waters'. No host could miss that going on. If a guest sniffs everyone will see, but if he doesn't the lure of this new odoriferous sensory environment will come close to driving him nuts. What most of us do is compromise when we go out to dinner, pretending not to sniff, acting as if we had not the slightest interest in inspecting through our nose the properties of the food placed before us, yet still taking care to hold our servings for just that extra moment before taking it in, just long enough to casually, hopefully with no one looking, tense our muscles to create that 20 mile per hour pull needed to bring the faint but oh-so-tempting odor molecules in.

Two species of living fungi on our cheese. Opposite: The beaded penicillin in a vein of Stilton. Below: Fern-shaped penicillin species in Cheddar. Both secrete poison to kill competing microbes — as do their medically useful brethren

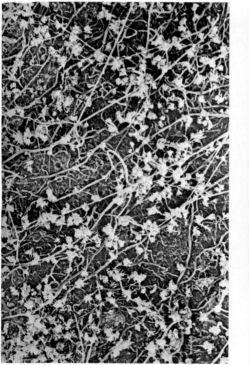

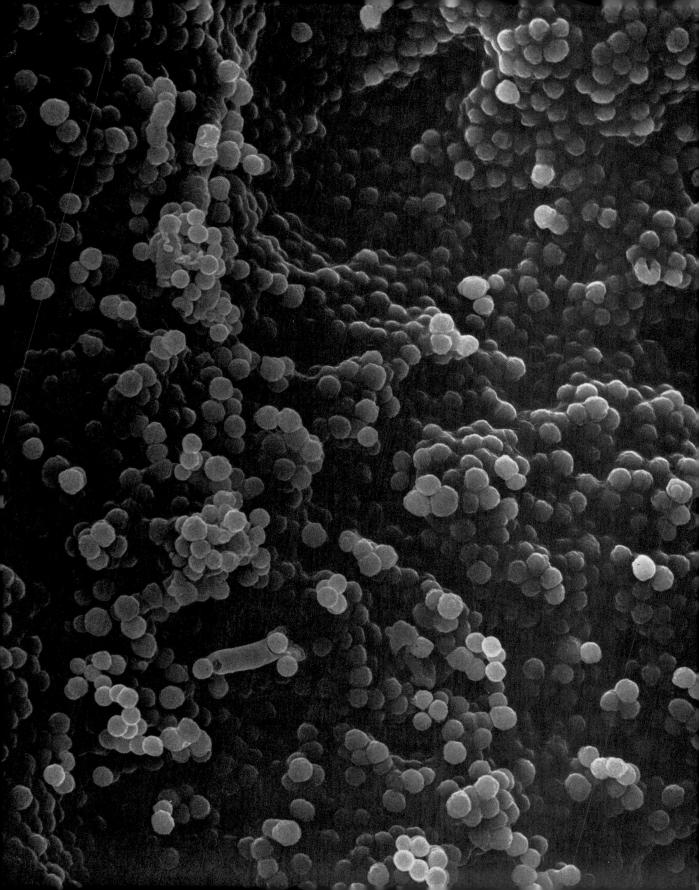

Salt. Pure sodium and chlorine ions will build up a perfect cube, but in table salt (seen here) impurities produce this extraordinary fleet of interlinked, only partially developed blocks. They are rigid enough to survive in this shape without breaking when shaken onto our food

Guests who have no shame about taking a full head-lowered, nostril-gaping sniff of the food, are also likely to take little time in creating a mess around their plate. There will be pulverized bread crumbs, slopped gravy, and what seem to be piles of errantly poured salt everywhere. The salt spillage at least is not an entire waste, as it is an excellent means of keeping down the bacteria which will be on and around your guest's serving of food. This unwitting devastation, going on unseen beneath the cheerful toast-making and more food guzzling overhead, is a multi-stage process. The first thing bacteria resting on our food do when surrounded by salt descended from above, is open nozzles in their body walls and sluice in plenty of water to try to dilute the stinging salt. On the plate of a delicate eater, who only lightly salts his food, the nozzles will soon close and the process will end there. But on the plate of a wild salt-pourer, eye on the next serving even as he still holds the salt shaker upside-down over his plate, the bacteria will have to keep on sluicing. After perhaps half a minute they become so filled up with cooling water that they resemble men in rubber diving suits that have swollen out Michelin-man style due to having too much water pumped in. On the plate the outcome is inevitable: let there be one more sprinkling of salt, and so one more reflex intake of water, and the whole creature will burst, leaving only futilely flapping body remnants on the food to mark where it had been. Spills of wine have the same result, but in the opposite way, drying the microbes they hit into hard rubbery disks, coagulated solid.

In the eating rush there is also likely to be a certain amount of spillage higher up than the plate, on the eater's face itself. Forks and spoons are so small, servings are so large, and anyway there are all those awkwardly shaped objects that have to be lifted up and stuffed in with an unlooking hand. Grease ends up on cheek and jowl, while gravy dribbles run down the chin. Since these substances were what provoked the urge to sniff when they were in the food, they will equally provoke a sniff or snort in their secondary resting place here. Garlic is especially potent, for the crushed cloves that were on the French bread contained alliin molecules, and while these by themselves are relatively large objects, with any noxious sulfur rings they contain ensconced well within, the garlic on the bread also unfortunately contains an enzyme ideally suited for releasing the sulfur segments from that alliin. Sticking to a smear of grease on the face or lips the enzyme will work its way through the alliin, setting loose the sulfur segments fast. These are not only odoriferous, but small, which means that they can easily take off from the perspiring face they've been created on, fly loose, and wobble across the room so someone else gets to share their distinctive smell. Indeed, if the emitter of this garlic has sufficiently numerous smears of grease on his face, a good number of alliin complexes will be preferentially absorbed on these grease flats, so letting the bespattered one chomp on in peace, well protected by this individualized gas mask.

But in time all good things must end. There comes a moment when the eater has to take a pause, lifting his coated face, wiping his greasy chin, and meekly ask for permission to depart to what the mediaevals called the *necessarium*, and we call 'the little room' or 'the little boy's room' or just 'the, uh . . . ' followed by a weak smile: the toilet. Our interest in accompanying him there, as he staggers down the house's corridors, peering into broom closets and whatnot to find the necessary chamber, will not be a prurient one. Rather it must be described because what a guest does in there, especially a guest who has a cold, is a matter of utmost microbiological concern for the family members who have to share dinner with him subsequently, and also live in the house after he has gone.

As a toilet flushes normally most of the water and contents go swirling down the drain, but because of all that swirling a certain aereated froth is momentarily created on the topmost layer of the water. It's only a few hundredths of an inch thick, but precisely because it is so thin it's not going to stay where it's created for long. This flush-induced froth separates off from the rest of the water as it goes down, hovers briefly in the air and then goes soaring up. It rises in a fine aerosol mist, far too diffuse to see or feel, but an estimated five to ten billion individual droplets of water are sent up in it. Lowering the toilet lid during the flush reduces their numbers, but only a bit.

The droplets in the haze are so small that gravity has very little effect in pulling them down. On the scale of the water droplets individual air molecules, as we've seen, are the size of flecks of gravel, and with the air molecules battering against the droplet from all sides, including under-neath, there's enough force to send the aerosol rising high within the room. The slightest air movement from the rest of the house will increase their flight — a door opening in the hall, for example.

Most of the aerosol droplets are simply pure water, but as the inside of a used toilet bowl is not the most sanitary of places, a good number of the droplets are not the most sanitary of objects either. The pure droplets will quickly puff apart and evaporate, but the ones with bacteria or viruses in them will form a minute ball around that living microbe core and stay intact in the air. These bacteria, as opposed to many of the harmless ones in the house, are pathogenic, i.e. capable of causing disease. They have just emerged from the lower digestive tract. Indeed human faeces contain a high percentage of viruses or live bacteria or bacterial corpses, and it is from this surface that the bacteria in the spray arise. Measurements show there to be between 60,000 and 500,000 of these pathogenic droplets after an ordinary flush.

Even though they are light and buffeted up by the air, these micro-droplets will in time settle down. Some start landing only a few minutes after the flush that launched them up, but most take over an hour, and a few will still be floating ever so slowly downward the next day. The microbes are cushioned by the water in the droplets when they land, and so almost always survive their touchdown intact. They nestle on the floor and cabinets, on the sink, toothbrush and wall. Some land on the doorknob and even the light fixture, and almost all have no problem going into suspended animation and surviving for up to eleven days as the transporting water bubble around them dries away.

Which takes us back to the dinner guest who has excused himself. Since he has a cold, he's also likely to have a slightly miscalibrated digestive system. The bacteria that are the cause of that join the viruses that have given him the cold, and all spray out in the air along with the general aerosol mist when he flushes. They will land all over the place just like the others. But while the host couple is likely to be somewhat immune to their own germs, they're less likely to be immune to this new contribution. Not only will they get it when they enter in the next day or two, but the flushing guest will also infect his outer surface from the haze as he walks out. Let him tap your arm while telling an anecdote, or worse, let him finger one of the serving spoons as he takes his second or third helpings, and what he let loose up there will soon be established on you.

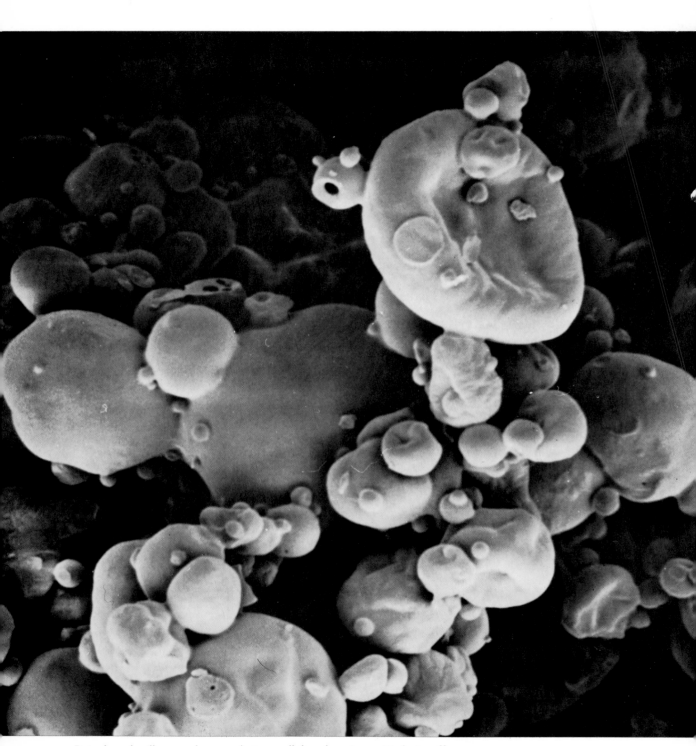

Powdered milk granules, ready to swell then burst apart in hot coffee

DINNER CONTINUES

Back in the dining room there's a sight to see. As the eaters sit fidgeting gently around the table, a jet of carpet dust, pollen and fungus spores is being sucked up from the floor and shot at high speed into their noses. This seems odd. How can it be happening in a dining room where not a single loose air nozzle or vacuum exhaust is in sight?

The cause is hot air. Human bodies produce heat, 98.6° F on the inside, a little bit less than that but still a pretty high temperture on the outside. This heat acts on the air just beyond the skin, and in heating the air makes it less dense. Approximately 2 gallons of air trapped between clothes and skin (1.6 square meters of clothed area times 5 mm of air caught underneath) are heated this way. Being less dense the heated air becomes buoyant, and, just like the great quantities of air heated by a propane torch in a hot air balloon, has a tendency to float up.

Down on the toes the heated air just separates off from the shoe and rises up in individual capsules, like the spray from an upside down miniature rainfall. That's an effect perhaps of aeronautical interest, but it doesn't amount to much, as the heated air capsules float away from each other and break apart even before they reach the level of the table. What's more important is what happens to the heated air produced at the ankles. As it wobbles up it doesn't spread out into the cool general room air, but finds itself in another gust of warm air, this time produced by the hairy calf, under its synthetic stockings, a little further up the leg. The ankle-produced air joins with the calf air, and wrapping tight around the leg the two join forces to continue their levitationary journey up.

Heat-flow photograph of a man in profile, showing column of warm air rising from around the body

Since they get joined by regular pulses of hot air as they move up the leg, the initial slithering airstream at the ankles turns into a higher speed gush by the time the level of the knee is reached. A stray amount puffs off blindly at the bend there, but most stays, and the steadily growing stream trundles along the thigh — under skirt or over trouser makes little difference — takes the second bend at the waist, and then is ready for the real sprint. Speed down at the heel and ankles had been 11 centimeters per second; here at the waist it's 25 centimeters per second and rising. The belly-circling ring of air accelerates, whipping over vests, corsets, petticoats or daringly bare midriffs, and makes it up to the level of the chest. Once again there's a certain separation: some pushing off horizontal and gliding out over the table; the rest, the majority, continuing straight up.

Two points are worth observing about what comes next. Up till now there had been no blunt obstacles to stop the upward bobbing hot airstream. Even a sprawling belly could be circumnavigated. But once the chest is reached there are some obstacles in store. The part of the stream that floats into the armpits gets blocked there, as does whatever portion rushes up to the earlobes. These are called stagnation points, and as a reconsideration of human sculpture will make clear, there's only one more protruding obstacle in addition to armpit and earlobe to create another stagnation point. In some members of our species this protrusion is of more noble proportions than in others, but in all cases it is there, and the air piles into it. This is the nose, and the reason we singled it out is that while armpits and earlobes don't do anything more with their heated arrivals than block them, the noble nose, at least in non-vegetative specimens, is also going to breathe it in.

Whence our second addendum. When the first air flow began at the heel and ankle, its disappearance upward produced a partial vacuum, and partial vacuums an inch above the floor are ideal for picking up dust. Into

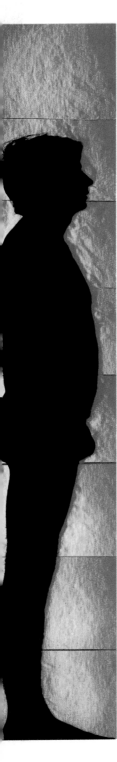

the staging area by the ankles the dirt gets lifted up, and with the next puff of heated air setting off it goes into the enveloping stream and with it travels up the body. Everything that was on the floor goes up: grit, pollen, fungus spores, asbestos particles, mite faeces, mite corpses, sweat residues, food crumb fragments – the lot. It's a mighty journey for those small items, akin on their scale to a large brick house being rolled all the way across America by a storm. The same happens to a minor extent whenever you put your hand out. Because of the heat it produces it acts like a snorkel tube. Dust on the table, spores from the cheese, salt fragments; all roll upwards, in this case along the fingers, hand and arm, and arrive seconds later at the neck to join the main stream. The portion that goes to the ears sticks in place there, giving justice to the eternal calls by observant mothers to wash in that apparently out of the way place. The part that goes to the stagnation point of the nose piles on in there – an estimated ten per cent of the air you breath when sitting or standing still has come bobbing and rolling up your body from the floor this way. But most of the floor-originating air flows up past the face – streaming within an inch of the eyeball – without stopping. It rises in the air till it's two feet over the head of each person sitting at the table, and there it spreads: an invisible geysering halo, for each and every one.

Let there be a clap of thunder, and the religious tableau would be complete. The storm which has been approaching all day is here.

If it's stormy, there's going to be wind. The thing to remember about winds is that they are sucked as much as blown. The breeze against your face as you walk on a sunny afternoon is not there only because a great mass of air in front of you is pushing itself towards you: air alone does not have the wherewithal to push. In addition a segment of distant air somewhere perhaps miles behind you, air with far more empty vacuum gaps in it than normal air, is also pulling the air in front of you back.

This can make quite a ruckus. There are about 20 million pounds of air in a column in the atmosphere for each one of us, enough to squeeze down with a force of about 15 pounds on each square inch of hand, head, foot and nose. (The reason we don't shrivel up from the press into a hollow prune-like shape is that some of that personalized 20 million pounds seeps inside and just as strenuously pushes out.) When this bulk of air starts moving it takes things with it. Some of the items in the air, such as the many million of microscopic creatures that are perpetually floating in a haze above us, do not have much effect on us however strenuously they are pushed around up there. But other items, such as the fragments of water bubbles at loose in the air, do something more. The winds high in a cluster of thunderstorm clouds can hold 500,000 tons of water and ice up there. That ice makes the air in the clouds colder and heavier than normal air, so down it falls, sinking earthward on whatever poor city happens to be underneath, a chill vertical wind that doesn't stop when it hits the ground, but continues bouncing outward, like a stream of honey spreading out on a dish as it's poured, gusting around everything it hits. For we mortals stuck on the ground underneath, the effect is what the ancient Romans called a *ventus*, a term which has persisted with little change in our 'ventilate', and which via a transformation through the Old High German *wint*, is what we now call a wind.

When the storm is directly overhead there will be enough of an aerial vacuum pull to make your house begin to swell. The windows bulge ever so slightly out, the walls bend out, and the air inside, feeling this irresistible pull, tries to escape in any way it can. If the storm is mild the air from under your dining table and over the steaming stew will merely end up floating up the stairs and out any upstairs windows which are ajar, or even through micro-cracks in the walls. If the storm is stronger though, and all the doors and windows are closed tight, then the air won't be able to get out, and will push and strain against the outer walls. It's like trying to carry on your dinner party inside a swelling balloon, and the resultant feeling of a clogged head that many people report is only what might be expected in such a setting.

There is a third stage, but it is luckily one very few of us have to experience. Swelling balloons all burst in time, and so do swelling houses. If the storm is really a big one, and it happens to be stationed directly overhead, then the suction upwards will be up to 1½ pounds per square inch. On a roof 10 yards by 6 yards there are 77,000 square inches, and the pull upward will be about 115,000 pounds. That's the weight of 5 large trucks. The result is that the roof flies off, exploded upward by the pull. Any sudden large noise can produce the same pressure pull as a storm, which is why artillery soldiers are advised to leave their mouths open when a shell is being fired: it keeps the air inside them from picking on the eardrum or eyes as the only way out.

Going out in a storm allows you to experience another interesting effect due to the wind. For this adventure you need a partner. Your gravy-smeared guest going out with you to help roll up the car windows will do fine. If in the race across the lawn to the car you say something to your guest he's likely not to be able to hear you. The reason is not that the wind has blown your words back at you. It's far too weak for that. When you speak the words always come out at Mach 1, or 740 miles per hour – the speed of sound – and so a wind of even 50 mph, that's a powerful Force 9 gale, will slow it down by less than ten per cent. What's happening instead is that the air at shoulder height and higher is getting pulled up more easily by the voracious storm than the stuff lower down, which is held back because it's knocking against the ground.

When you speak in the race across the lawn accordingly the sound waves of your voice get tipped up by this air speed imbalance. They reach the vicinity of your friend without being diminished, but by the time they get there they've been tipped so much that they're floating several inches above his head. In a very strong storm even yelling won't work, because your sound waves will be tilted up even more sharply. Supplying your partner with stilts would help a bit, while parking him on the roof of your house would do even better: your voice in a storm can often be heard quite clearly up on the roof. Of course if engaging in the latter exercise, your yells might soon be transformed to ones of forced explanation to any police officers who happen to drive by: the spectacle of a man screaming to himself in a storm, and saying it's to contact a dinner guest sitting on his house roof, is enough to rouse suspicion from those uninitiated in the peculiar nature of the wind.

Along with the wind, the rain. Rain is the result of a storm cloud self-destructing. In a normal cloud the vapor that makes it white and fluffy is too diffuse, too light, to do anything more than bob gently up and down in the air. That's why normal clouds just plod along without problem through the sky. In a storm, though, the normal cloud transforms. It's thicker than before, so that little light can get through, and in extreme cases the once white cloud will appear black. And also inside it the water vapor, all those thousands of tons of the stuff kept up there by the winds, is doing something it wasn't doing before. In ordinary clouds the water vapor is wrapped around certain very small chunks of matter that happen to be caught in the cloud. A few thousand water molecules on such a core will not be much of a problem, and the cloud will still continue to float. Even a few million water molecules on each core will not make any change. But in a storm cloud the water keeps on wrapping around those cores, the globs that get formed become larger, and ungainlier, and heavier – and soon they are too heavy to stay up in the cloud at all. They fall, fragments of the cloud breaking loose from their usual resting place in the air, and we call their descent rain.

In thin and broad clouds (forming strata, or layers, and so called stratus clouds) the rainfall can go on for hours, since water vapor in little microscopic puffs keeps on floating back up from the street or soil where it hits to replace the larger visible droplets coming down. In clouds that are narrow and tall, however (accumulating in vertical chunks and so called cumulus clouds), so much rain is formed inside, and so many droplets come bucketing down, that there's no way enough rebounding loose water vapor can float back up against the onslaught. Those clouds accordingly rain hard, but quickly rain themselves out. Such clouds being rare in northern cities generally, it is the unfortunate fate of residents of those misplaced habitats to suffer the steady and self-perpetuating drip drip drip of the gentler stratus type.

But where do the convenient core materials come from that allow the vapor in a cloud to build up sufficiently weighty drops to fall down? As it turns out there are many substances at hand in the atmosphere for this important role. Some is just ordinary dust from the neighborhood; some is errantly floating sand that's ended up here from distant deserts. But perhaps the most unusual of them all are those core particles which came not from the neighborhood, not even just from a desert on the other side of the globe, but from very much farther away indeed: particles which just five weeks before were skimming quite contentedly through the blackness of empty space. These are the micro-meteorites.

The meteorites had been out in space for quite a long time, 4,700,000,000 years by current estimates, which is just a little bit longer than the earth we live on has been here. The meteorites in fact seem to be fragments of the original substance the inner solar system was made of,

A plummeting drop in a puddle, rebounding up in a geyser and forming a perfect sphere

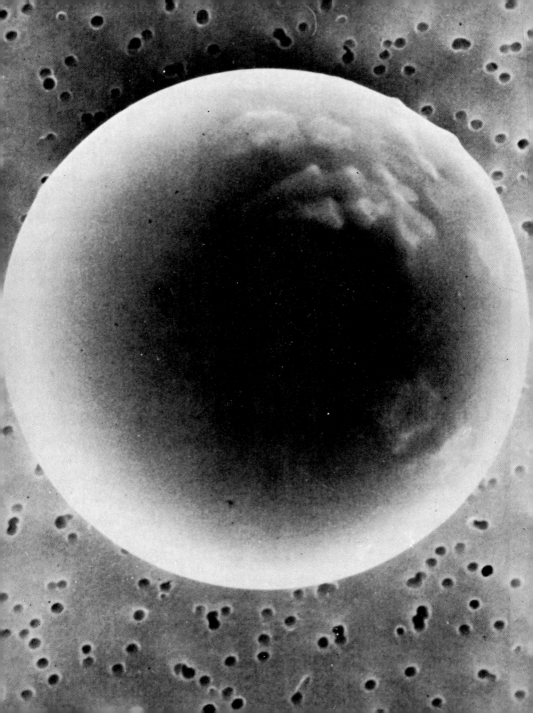

and while their brethren were busily crunching together to make the earth, Venus and Mars, these little fragments were passed over, forgotten about, and left outside. Some were metal, and would have gone into the ore deposits that ended up as your belt buckles, doorknobs, calculator innards or kitchen whisk; others were stony, and would have ended up as pebbles in your garden, parts of Mt Everest, or other structures we expect stones to be found in. Left over at the earth's creation, though, they had no choice but just to orbit the sun in slightly eccentric paths, sometimes nearing the planets, but mostly off by themselves.

Quite by chance the earth sometimes collides into those lonesome fragments, bringing them back after all this time to the planet they were so close to joining at the creation. For the larger meteorites it is a flamboyant homecoming, but a brief one. They hit the atmosphere faster than the Space Shuttle does on re-entry, and lacking that vehicle's ceramic tiles on the bottom, they do not get to share that vehicle's gentle descent. Instead they heat to a temperature where metal or rock turns into flame, and burn up or even boil away immediately upon entry, producing what we sometimes see on clear nights as shooting stars. Traveling at a relative velocity of 47,000 miles per hour, these 'shooting stars' are probably the fastest moving object in the earth's neighborhood you will ever observe. A few of the largest ones survive in this flaming rush all the way to the surface: these get a name change, and become meteorites. (For many years more were found in Kansas than seemed fair; the puzzle of why they should have this predilection for the American Midwest was solved when someone realized that there were probably no more in Kansas than elsewhere, merely that in a flat terrain of corn and wheat a sizzling boulder dropped from the sky is easily picked out.)

For the smaller meteorites arrival back on earth is less transitory, if also less flamboyant. They get braked by the earth's atmosphere before they have time to heat, almost as if they had sensible little parachutes on. In a few minutes they're down from 47,000 miles per hour to a more prosaic 1½ miles per hour, and a few minutes later they're down to only a few hundred feet per hour. At that rate the earth's atmosphere seems viscous, like thick shampoo would to a pearl, and they take about four weeks to seep all the way down. The numbers of micrometeorites are surprising. The total mass entering our atmosphere each day is estimated at over 6,000 pounds, divided among many, many trillions of individual particles. If there's nothing in the way they'll land on the surface; an extremely thin film of them can be detected with the right equipment having landed straight on your house roof, and even on your windows. But if there's the slightest updraft of air shimmying in their way, as might be produced by a cloud in the making, then the 0.002 mm meteorites will slow and come to a stop, still many thousands of feet up in the air. Ending up in the white moisture clouds so precariously bobbing above our homes, they serve as convenient hard cores for the moisture to condense into a rain droplet on. Only when the mass of enveloping water is large enough and heavy enough — only when the cloud is turning into a storm cloud — will they fall. Hold your hand out to catch a few raindrops, and a particle older than the earth, which has traveled trillions of miles and just arrived from outer space a few weeks before, will be in your palm.

As this laboriously constructed rain lands, it sends the inhabitants of your garden into action. Earthworms will start wriggling at top speed to the surface to avoid drowning as their open-ended subterranean tunnels fill up. Ants start running around their underground nests in search of the tender infant ants, and, picking them up in their jaws, climb along the ramps and passageways in the nest to bring them to a safer level. In a house threatened by torrential rain this safer level would be up; but in an underground ant nest getting soaked from the top this safer level is down. Once the babies are downstairs the ants will turn around and start banging their own heads against the walls. This is not some excess stress ritual, satisfying as for humans because of the pleasure ensuing when they stop, but rather a way of compacting the sand and dirt particles that make up the walls. In lieu of thumbs it's the most efficient way the ants have. There is minimal brain in those skulls to get mashed around from the impacts, and the only signs of tiredness are the quicker breathing, and the condensed moisture beading up on the ants' oily bodies as they so happily smash down there below the storm.

And with the wind and rain comes lightning. When the storm was still 20 miles distant, the steady trickle of invisible 'electric rain' from the sky that we met in Chapter One started to slow. There was a moment when it came to a stop, and then, as the storm got closer, the electric rain over your house and garden started to come on again, but this time ridiculously uneven. Instead of falling down from the sky it starts streaming up from the ground, rising off the left-out lawn chair; then once it's done that for a bit it switches and goes back down, only stronger than usual, and again only briefly till it reverses once more. To get such reversals there has to be something very powerful in those clouds capable of attracting all those many electrical particles up; a lightning bolt in the making will do fine. The thundercloud becomes a floating electrostatic generator overhead, separating out positive and negative charges inside until finally the electric field between one of the charged parts of the cloud and the earth grows so strong that the air in between can't take it. There will be a gush of electricity – what we interpret as a lightning bolt on its way.

When a bolt of lightning starts to fall down from a cloud to earth it does not take a simple straight line. This can be seen from what we remember of looking at lightning bolts: they're never straight, like a fireman's pole, but rather ragged and uneven, as if they hadn't so much plummeted out of the sky as staggered down. The evidence of high speed cameras shows what happens. First the lightning bolt drops a hesitant little probing length from the cloud base, only 20 yards long or so, and barely wider than a finger of your hand. That slender probe just hangs there, quite still, for a few dozen microseconds, before it builds up force and, after poking and fussing about sideways a bit to find the best next path, spurts out another 20-yard extension at a slight angle to the first, so making the whole fiery pole from the cloud now 40 yards long. There's another pause, then another 20-yard length is added on, and so in such hesitant and jerky steps the initial stroke makes it almost all the way to the ground.

While it looks a bit silly in such slow-motion displays, an even closer observation, using ultra high speed computer processed equipment, reveals that the descending bolt is pretty vicious in those moments it makes its growth. Some of the lightning bolt lengthening comes from charged particles gushing down inside it, like water falling down a drain. Some is just torn out of the air in its way; often its frying ultra-violet light burns the space in front of it, turning the air into ragged charged particles which accelerate over to the bolt and stick on too. Loose cosmic ray particles originally from distant galaxies that happen to be nearby get pulled in and pasted on to the growth, and it's between them all that the staggering bolt fits on the length it will need to approach the ground.

DINNER CONTINUES

Once the descending charge is within a few dozen yards of the ground, say about twice the height of your house, the lightning bolt is ready for the real fireworks. (Despite all these stops and starts it has only taken about $\frac{1}{10,000}$ of a second so far.) It seems that one reason lightning exists is to balance out the downward electric field that produced the movement of charged particles as 'electric rain' whenever the weather is fine. It replenishes the lost charge we get to experience every time we open the window on a sunny morning. If the descending lightning bolt just stopped above the ground, or went ahead and hit it, that function would not be fulfilled. If anything, matters would be worse, with even more charges having fallen down. What the fresh lightning has to do instead is return back skyward once it's close to the ground, and make this return journey loaded with enough excess electricity to counterbalance the falling of ordinary current from the invisible electric rain.

That it does. Back up to the cloud a return stroke explodes. It follows the same channel cut out in the sky that the original descending one used, but there the resemblance ends. The downward initial stroke travelled a few tens of thousands of miles per hour, about the speed of an ordinary re-entering spacecraft. The upward return stroke however moves at 60 million mph — almost a tenth of the fastest speed obtainable in the universe. This upward stroke takes so much electricity with it that it glows at a temperature of 25,000 C. That is hot; the surface of the sun is only 5,000 C. What you see of a lightning bolt is the second part of it, which doesn't fall from a cloud, but starts on the ground and streaks up to balance the charge.

Because the lightning bolt is moving so fast and with such terrific force, the air immediately surrounding it is thrown outward; then the air around that wrapping layer is itself thrown outward, and after a series of such aerial pummels a final layer of outward-thrown air reaches where you are standing: it's what we know as thunder. (An earlier explanation of this phenomenon exists in our word 'Thursday' — a corruption of Thor's Day, named in honor of the Norse God of thunder, the red-bearded Thor.) At a distance of two miles thunder sounds like a thud followed by a rumble. If you are ever unfortunate enough to be less than 50 yards away from an impacting lightning strike, the blast of expanding air will sound first as a sharp click, and then turn into a crack like an extremely large bullwhip being snapped beside your head. In this case the whip is the several-mile-long lightning bolt, and the cases of temporary deafness reported after such an encounter are understandable.

Where the initial bolt hits the ground and the return stroke takes off again there's a terrific explosion. Part of the bolt sinks down into the soil, and for any bacteria or worms that happen to be stationed there the results are certain to be fatal. Lightning hits so hard that sprays of calcium have been detected squirting up where it touches, and that calcium can only have come from shattered rocks or dissolved small creatures. It's also likely that the lightning bolts you notice during a storm are zapping some of the nutrient broth on the ground where it hits into amino acids and other precursors of life. Unfortunately little is likely to come out of that creational endeavor, for the first fungi or bacteria recolonizing the charred region later will gobble up that new substance, and so destroy the chance of our ever seeing what new life forms may have grown out of it. There are an estimated 1,800 lightning storms in action on our planet at every moment, so this happens a lot.

Such inhospitality to newcomers seems to be the standard practice on our planet. Every living creature today – you, your dinner guests, cows, whales, daffodils and mites – has amino acids which polarize light in a leftward direction. By simple chance numerous proto-life forms must have started to develop in the earth's primitive ocean which were mirror images of that and polarized light in a rightward direction. The fact that no descendants of such creatures have been found, anywhere, shows that our ancestors, bacterial or otherwise, destroyed these newcomers before they could develop enough to fight back.

If the lightning hits a tree in your garden, the effect on the tree will depend on how long it has been raining. If the tree is wet right to the base, the lightning will land on its topmost branches, and follow the water film down the trunk to the ground. If the tree is not wet all the way down however, then the lightning that lands is much more destructive. It goes down the outside of the branches all right, and even starts down the outside of the trunk, but where the water runs out it will jump to the sugary sap on the inside of the tree and try to follow that down to the ground. Now sap is held in narrow channels inside a tree, and when the lightning five times hotter than the sun hits the sap it vaporizes, and not

having any place to expand, it blows the tree apart. A slight film of caramel on the stump – from the flash-heated sugar in the sap – is the only giveaway. Oak trees suffer from these explosions more than others, for they have roughened barks, and for a given duration of rain, water will trickle less far down an oak than down a smoother girthed tree.

Standing under a tree during a thunderstorm is counter-recommended not just because of the explosion. Even if the lightning bolt does get all the way down the trunk it is then likely to move out horizontally along the thickest roots. As these are often close to the surface, someone standing upon them will get electrocuted by a charge entering one foot, then travelling up the leg, across the body, and down the other leg, to exit into another near-surface root. This accounts for many deaths every year, especially on golf courses, though there have been fortuitous escapes. There have been several cases where men were struck by lightning this way and survived, with only a melted nylon zipper in their trousers to reveal how the current managed to pass from one leg to the other without doing harm.

Cars are safe in a lightning storm because the current of an impacting bolt will roll harmlessly around in the metal body, and so long as no one within decides to conduct an experiment and touch the surface it will stay there till it dies out. The same goes for commercial jetliners, which are regularly struck by lightning in flight. The estimated figure is over 100 strikes on any commercial jet each year. There too the current rolls harmlessly off, with the only sign of this aerial electrical assault being the cluster of pitted spots you can see on the nose or wingtip if you look close the next time you're boarding.

A lightning strike on your home will often travel through harmlessly in the same way, going from TV aerial down through the wall circuitry and so to the ground. Still a lightning rod is a sensible investment, as the argument used against it by the good pastors of Philadelphia when Benjamin Franklin invented the first one in 1752, that protection against God's wrathful electric will was impious, finds few backers today. Only residents of homes guarded by them can truly savor the pleasant game of seeing how close a lightning storm is, and whether it's moving towards you. The rules are simple. Count the time from lightning flash to the first thunder boom, then divide by five, and you'll get the distance of the

DINNER CONTINUES

lightning bolt in miles. (That's because thunder, like all sound, takes about 5 seconds to travel each mile.) Count the time from the first thunder boom to its last rumble, and by similar mental arithmetic you can tell how high the cloud delivering the lightning is: for a stroke nearby a five-second rumble will mean that it's one mile high. If the rumble gets shorter, then the cloud is coming lower. If the interval between flash and boom decreases, then the stroke itself is coming closer. The winner is the one who first hears the click of a bolt directly overhead – though by this stage it's probably best not to be standing next to any exposed glass windows.

Back in the dining room – the two men having survived their stormy errand and dried off sufficiently to resume their social roles – the table is cleared to make way for dessert. It is a crucial moment. During the main course there was plenty of food to chew, and the guests were kept busy, pounding away at the standard chew rate of 100 tooth blows per minute. This soothed and placated them, quite likely inducing, as food technicians studying chewing behavior report, 'decreases in fidgeting activities such as finger tapping, leg swinging, adjusting hair . . .' But without the food there, especially with the excited interruption of the storm, they might now go berserk. Something is needed to placate them, and fast. As the bread has already been used up, there is no choice but to let them eat cake.

If the serving of cake you present to your guests for dessert were divided into its components, the result would be a bowl of water with globs of fat floating in it. There would also be a coating of sugar, and, underneath the wreckage, a certain residue of flour. All cakes are like this, or at least all cakes you get from commercial bakeries are, for such cakes are not really food but just a way of taking ordinary water, mixing it with low-cost fats, and then disguising the result so it can be sold at a profit of several hundred per cent over raw material cost.

The process begins with cake manufacturers collecting the lowest cost fats they can. Olive oil is never used, because that substance has a pleasing texture and satisfying smell, and so can be sold at a high price on its own. The stuff shipped in tanker loads to commercial cake factories is less pristine in origin. There's usually a good deal of lard, a good deal of oil from compressed and over-aged fish corpses, and perhaps a little palm oil too. The fats and oils are mixed together, cooled until they're fairly stiff, and then the air is blasted in.

Blades the size of large airplane propellers force air into the fat chamber. Because lard and fish fat are so sticky, the air doesn't carry on through but gets caught up inside the mix, separating into little bubbles as it gets absorbed. When the fans are turned off what's left is a giant block of aerated pig and fish fat. This is not a substance that could be chopped into segments and successfully marketed as cake.

The first thing that has to be done to this cake precursor is to find some way of making it thinner. It's too heavy as it is, and anyway there's another advantage to making it thinner. The lard and fish oil were not very expensive to start with, but if they could be made to go twice as far, then whatever profit there might be in store could also be spread by two. For this diluting a substance known as GMS – glycerol monostearate – is mixed into the fat. It is chemically very similar to soap. With the GMS in there any water hosed into the room where the fat is will not bead up on the surface, as water tends to do on normal fats, but will instead be pulled in by the soapy GMS. GMS is so good at pulling water into fat that hundreds of gallons of water can be sprayed in for each ton of mixed fat waiting in the storage rooms to become a cake. The compact fat wedge swells as the water comes in, it grows and stretches and because the GMS does so good a job it ends up as a super-wedge, twice the volume of the original.

Once the water is there something else has to be added to the mix: sugar. Crates and crates of refined white sugar are dumped on, stirred in, and left to spread. Since sugar dissolves in water it is taken up throughout the volume of the fat, wherever the GMS-assisted water has itself managed to spread. Sugar helps the fat and water concoction smell sweet, but mostly it's there to add weight – important as the aerated water and fat mix is now too light! Sugar turns out to be one of the most inexpensive substances of substantial weight that is safe enough to be added to food. Gravel, logs or cement weigh more, but happen to be fatal if ingested in quantity; flour, protein and other nutrients are certainly safe to add to food, but they're not as dense, and don't weigh as much. Sugar is the only substance that falls between the two, neatly replacing weight the water took away. It sounds roundabout, but has its logic.

By this stage the proto-cake has 90 per cent of the ingredients the finished product you eat will have. There's pig fat, oil from crushed fish, lots of water, and lots of sugar. It's not a very palatable object, being a pasty grey in color, and oily as you might imagine a great hunk of old fats would feel, but with a bit more transformation all those irritating lacks can be taken care of. First some flour is added. As it's masked by all the fat, sugar, and water, there's no need for an especially high grade of flour to be used. Often it's the reject from bread-making factories. Even so it's expensive stuff, or at least when compared to plain water and aerated fat it's expensive, which is why only a small amount is used. All it has to do is provide a thin filler to go into some of the fat sheets that have wrapped around the air spaces, and an addition equal to four or five per cent of the total cake weight is usually enough for that. Sometimes the flour is dispensed with altogether and simple cellulose derivatives − ground-up wood chips − are used in its place. These have zero nutritional value, but fill the fat membranes almost as well. Meringues especially are likely to get this substitute.

The GMS that was originally added to hold the water inside the fat now has another role to play. Left to itself the flour added in might crumble into small clots, and so give the cake lumps. The soapy GMS oozes around those flour pieces though before they can clot up. That keeps the cake-to-be even and free from lumps.

Only a few faults are left now. The cake still looks pretty bad, and so is coated with coal tar colorants; it also tastes intensely objectionable − soapy, oily and greasy despite all the sugar − and is injected with some flavor to make it palatable, usually one of several hundred strong synthetic flavorings on hand.

With all those additions since the first air treatment was given to the fat mass, the cake is likely to be getting pretty compact again. Baking sodas have to be added to get it to rise, to get the fat-covered air bubbles to grow. The cheapest baking sodas leave washing soda in the mix as a byproduct, and as this is itself pretty nasty tasting it is only used for cakes that will end up as chocolate − a flavor that can be made strong enough to cover almost anything left inside. This is a general rule. In almost every commercial food-making process, a batch that gets spoilt will be flavored with chocolate to get it through. For other flavors a slightly more

Delicate flowers of crystalized sugar, made visible in a polarized light micrograph. This is exactly the same as that sprinkled on your food

DINNER CONTINUES

expensive baking powder is used, which contains an acid capable of dissolving undesirable byproducts as it goes along. In both cases carbon dioxide is chemically generated inside the cake, and being caught by those flour-toughened fat membranes, swells them up. What started as a simple lard, fish oil and palm oil gloop, is now an epicure's delight, airy, light and tender to the touch.

There's a footnote on consumer psychology here. When the first home cake mixes using the GMS magic were marketed in the US, they did not sell well. Consumers felt that an amorphous substance you just added water to and baked could not a true cake be. They were right, but that wasn't the issue. The manufacturers might have said that at least the powder was better than what you get in a factory cake, but that did not seem an attractive point, and anyway it was probably best to keep quiet about what went on in the factories. It looked like the product would have to be withdrawn, until an ingenious advertising man got the idea of saying that a fresh egg had to be added to make the mixture work. It didn't have to be added, the GMS chemistry worked fine without it, but it gave the housewife users a feeling of being in control, of creating a natural product, of doing good for their family: sales of cake mix went up.

What the host serves the guests with the cake, the large dollops of ice cream, is an even greater case of manufacturer's ingenuity. It is not just water that's being sold at a whopping profit – though there's 30 per cent of ordinary tap water diluting your ice cream. Nor is it just a question of the fat (6 per cent of ice cream) or the sugar (7½ per cent). Rather as a standard text proudly puts it, ice cream is 'noteworthy as the only major food product in which air is the principal ingredient'. Fifty per cent of the ice cream you buy is nothing but empty, zero cost to the producer, air.

Ice cream does this by having a clever internal structure. Like cake it starts off by being a slab of fat pumped with air, but in ice cream the fat is toughened and made rubbery so that it can hold even greater volumes of air than in cake – an unpleasant process to observe, taking place as it does in cold rooms where the layers of freshly forming fat-bubble sludge are continually being scraped off the freezer wall and taken to another cold room to package. There are stories of the newly nascent ice cream falling to the floor as it's being scraped off; this is later collected and, as might be expected, flavored and sold as chocolate to mask the taste it picks up lying for hours on top of the metal floor grid, and under the workers' trudging boots.

To get the aerated fat mixture ('ice cream mass', the food engineers call it) into something resembling ice cream, it's not just enough to make it cold. Then you would have only a foamy cheap margarine. Something has to be done to cause the slab of fat membrane to flow, to make the mix go gooey, stick to the spoon, and do all the other things ice cream is expected to do. That something is a hypodermic injection of glue. It is produced from boiling down the parts of cattle and pigs that no one else will eat – udders, nose, tail, rectal skin – and once injected it flows and spreads over every membrane of fat in the ice cream. This glue is what makes the ice cream gooey. Let your guest stroke the broad bottom of his spoon against the ice cream, and what the glue network will do is vibrate like so many miniature rubber cords. The ice cream will merely shimmy, and the satisfying dreamy-eyed spoon play can go on.

Glue does more. When the carefully felt-out frozen object is brought to the mouth, it's the glue that ensures that the ice cream melts in a satisfyingly smooth way. This is essential. A food will only be accepted as of the mushy-sensual type if it immediately goes into viscous flow when pressed between the tongue and the roof of the mouth. Such viscous flow is ice cream's forte. The rubbery glue strands are so gluey that even the water mixed in during production got attached to them. When the mass was frozen, that water froze in place, spread out neatly in tiny crystals along the threads of glue. If the glue hadn't been there the water would have ended up as simple ice cubes. Tasting the result would have been like putting a load of ice cubes in your mouth. But with the glue-spaced ice crystals, putting the ice cream morsel into the heated mouth means that the tiny ice crystals only break loose and turn to water in commensurately tiny droplets, one by one. This produces a meltation so smooth, so satisfying, that particularly gluttonous guests can be observed spitting the partially molten ice cream back onto their spoon, so that it will refreeze and can then be slurped back inside their mouth to repeat the pleasure. Those neatly arranged ice crystals dripping off the glue tendrils are an irresistible treat.

Dinner over, last ice cream segment publicly slurped down, etiquette demands yet another stint with the guests in the living room. There will be sweaty bodies draping themselves over your sofa, bellowed demands for chocolate, for pretzels, for fruit and medicinally necessary fiber. Even this will not stuff them full, for soon out of their mouths, out through the space you can see where the sloppily masticated pretzel solids are not quite jammed solid to the top, there will be words: streams and gushes of meaningless words, drivel, nonsense upon stilts – and all yours until the guest gives up.

If the prospect is too upsetting, there are several possibilities for bringing the evening to a speedy end. One is assault with a blunt instrument, but this, though quick, can be messy. Far better is to suggest that your guests clamber up on a chair. As they stand there, groping on the top of the bookshelf for the non-existent hardback you say is there, their bodies leaning, precariously swaying, you do not even have to do anything as uncouth as push. All you have to do is fold your arms and wait: the wonderful world of statistics will, in time, do the rest.

In every country where measurements have been made, death by falling is the leading cause of accidental home death. The only exceptions are Sri Lanka, where falling into home wells is more common, and Japan, where death by suffocation in the home just edges out falls. In Scandinavia falls are virtually the only way you can expect a guest to die in your home, especially females: figures for that one category cover 94 per cent of all home deaths in Norway, and 82 per cent in Finland.

If waiting for a fall doesn't work, then other methods from the statistical repertoire can be tried. Suggestions that the guests clean or play with the household gun are especially fruitful in the US, where home deaths from gun accidents reach over 1,200 per year. In England there are too few guns for that to work (only 22 deaths yearly in the UK from home gun accidents), but poisoning and burning are fair bets, while if you could bring in a copious offering of nuts and other nibbles, death from suffocation will be surprisingly likely. About 350 people choke to death on their food each year in Britain at home, a figure which when adjusted for population is the same percentage level as the American home shooting one. If you are very patient you could ask the guests to re-park their car across the street over and over again until you say stop – each 780 million miles' driving gives a 50 per cent chance of a fatal accident – while if you are very, very, very patient you could just ask them to stand by the window and look up: the expected meteor impact rate has been estimated at one strike per human being per several million years.

Or you could just ask them to smoke.

Smoking one cigarette takes a statistical 1½ minutes off the smoker's life expectancy. This doesn't sound like much, until you realize that for a long-term several pack-a-day smoker it adds up to eight years less. That's promising. Unfortunately, smokers have the awkward habit of refusing to inhale what their cigarette produces themselves. Indeed it sometimes seems that they try to inhale almost nothing at all. Videos show that an ordinary cigarette smoker takes no more than 11 or so puffs in the duration of her smoke ('her' because most smokers in Britain and America are female). Each one only lasts for two seconds. The rest of the time, 300 seconds or more, while she talks, exhales, sips a drink, or just waves her arm to illustrate a point, the cigarette's combustion products are let loose to float across the room.

To see what's in this so generously released smoke, it helps to imagine a truly demented dinner guest wheeling in a shiny aluminum cauldron and a trolley of chemicals to do some magic tricks for her hosts. First she could drop some fine metal shavings in the cauldron, then pour in a bit of caustic soda to react with them and send up an exciting billowing cloud. Such a display might produce applause and even a cheer, but a chemically obsessed guest will not stop there. A squirt of paint stripper might go in, then some interesting methane (the gas given off by rotting swamps), and then some ammonia. Even the best-mannered host might be tempted to call a halt at this point, but before he could do so we have to imagine the guest, with a fiendish grin, whipping on a gas mask and uncorking even more containers, ones marked this time with the international poison symbol of a black skull and crossbones. In those go to the cauldron too: nitric oxide, formaldehyde and others. The hosts and other guests would have collapsed on the floor by now, the brandy spilling from their hands, but the chemical mixer we are imagining is not done. With a fiendish grin she prises open a flask that had been especially well covered and brings out vials of concentrated and quite deadly hydrogen cyanide.

Our actual guest, without benefit of cauldron, is pouring out exactly this mix of chemicals as she lights up a cigarette in your living room after

DINNER CONTINUES

dinner. There are reactive metal fragments in a burning cigarette, as well as ammonia fumes, the paint stripper chemical acetone, hydrogen sulfide, methane, hydrogen cyanide, nitric oxide, formaldehyde, mosaic virus, and over 1,000 other delights, ranging from irritants to poisons, nerve gases to mutagens, and lots and lots of carcinogens. The reason cigarettes are the source of so many unlovely substances is not that they're packed with them in the factory, but rather that the lit cigarette makes them itself. It's easy to forget just how hot the glowing tip of a cigarette can be. During inhalation it will reach 1,700° F, which is more than the temperature of a branding iron at work. That heat rips the tobacco and paper compounds into their constituent parts, and then, from those basic parts, builds them up again into the complex, and poisonous, chemicals we started with. It can do this because in addition to the heat a burning cigarette creates within itself boiling water, made when ordinary hydrogen and oxygen start to come together under the heat to form H_2O. That water superheats into steam, and so a steam distillation laboratory in your guest's very own cigarette is switched on.

It's not surprising that these chemicals arise. Tobacco plants evolved the ability to make nicotine – originally a nerve gas against insects – through simple transformation on the constituent chemicals in the leaf. With the heat and steam in a burning cigarette, a smoker can to some extent repeat these transformations and easily go on to create more. It's an ideal factory arrangement: high temperature blasts from the puff to start things off, then a cooling zone for steam to condense and help the chemicals form. And with the habit of brief puffs and long pauses there's even enough time for all the hundreds of necessary chemical syntheses to take place. The region in a cigarette where all this happens is called the pyrolosis zone, and you can see it as the dim glow just inside the cigarette from the red-hot tip.

The newly-created poison chemicals don't plop off in individual molecules once they're created. Rather, in the smoke stream you see slowly curling away, the chemicals clot together in what look like little tennis balls. They are extremely small, just ⅕ micron in diameter on average. It would take 10 rolled straight in a line to be as long as the pseudomonad bacteria we met on the kitchen table earlier. About 200 billion of these clotted chemical balls bounce out of a cigarette while it's smouldering between two puffs, a release rate of 6 billion per second. Being hollow the chemical balls float, which is why cigarette smoke goes up, not down. In those vast numbers few of the balls will be identical. Some might be especially rich in the embalming fluid formaldehyde that's synthesized in the hot cigarette; others will be built more of ammonia and acetone. Once your guest's cigarette has burnt to the halfway point, miniature tennis balls that are especially rich in hydrogen cyanide will be bouncing out.

In time all the chemical bubbles will fall, brought down more by air currents than gravity, and wherever they land they will stick. Hair is good, especially if it's been washed recently and so acquired a negative electric charge. Walls are excellent absorbers, while fabrics such as clothing and furniture upholstery are nice landing spots too. Emanations from the smoker herself are added. Several billion miniature acid, ammonia, cyanide and formaldehyde hollow spheres will spray out of the smoker's mouth as she exhales after a puff. These are almost the same as what came directly from the smouldering cigarette, but not quite. First of all these exhaled particles are bigger than the others, and so heavier, which means that they will be among the first to drizzle back down on you. They will have more mutagens, more corrosive acids, and, if the smoker is wont to exhale through her nose, they will also have an added amount of burnt sugars and mucus constituents (from her nasal lining) to join the exhaled spray of her friendly offerings on to you.

It is more than can be endured; the time has come to escape. For the well brought up host this will not mean an eyebrow-raised glance at the wristwatch, followed by a noticeable yawn; nor even a wild, inhuman growl from deep in the throat followed by a mad lunge and leap through the living room windows to the bliss of the muddy but mercifully uninhabited garden outside. The neighbors would comment, and word gets around. Rather the solution, one all but the saintliest among us have indulged in at one time or another, is to appear to stay plausibly in our seat, to even continue stirring the coffee or gently drum our fingers on the side-table, but to let the mind escape. It is time for musing, for reverie, for the living room to be left a million miles away. There are, after all, so many interesting things to hand to muse upon, and such models of great thinkers who have mused on them before . . . .

DINNER CONTINUES

Consider the simple stirring of the after-dinner coffee in its cup. As the spoon goes around, the coffee goes with it, and the edge of the fluid rises up by the cup's edge. On first reflection this seems simple: the cup is still, and the coffee is spinning around inside it. Like a fair-goer being thrown to the side of a spinning carnival ride, it has to go up. But how does the coffee know it's supposed to do this? By all rights, from the coffee's perspective it could just as well be said that the cup is still and the universe is rotating around it, just as when we're sitting in a stationary train we might think we're in motion when, as the station master knows, it's only the next car going off. And if the cup and coffee are stationary, and only the universe outside rotating, then there's no reason for the coffee to rise up. Who is the universal station master then?

This is not sophistry, but has perplexed the greatest scientists in history. Newton got around it by saying that such a cup was still by reference to the fixed space around it. His notion was that some absolute co-ordinate system is laid down on us from 'out there', letting us know when we're doing the spinning, and when it's the rest of the universe that's moving instead. This is *ad hoc*, and although most scientists accepted it because the great Newton had decreed it, Newton himself in private writings reported doubts. These doubts he was never able to resolve. (It is possible that this is what precipitated his various nervous collapses, abandonment of all physics, and flight into mystical speculation.) Only in the last century was the problem approached again, by the Austrian scientist and philosopher Ernst Mach (whose name survives in our shorthand for the speed of sound, Mach 1). He suggested that there was nothing from distant space or the distant stars which reached down to our spinning fluids here on earth and let them know when they were spinning; rather it had something to do with the nature of the fluids in the cup alone. What this something was, though, he could not specify. His musings had some fame but because they were so vague were generally ignored, until the best-known patent office clerk Switzerland has ever had took it up. Could the answer come by delving into what would happen if the perspective of *neither* the coffee nor the surrounding universe were to be preferred? The young patent clerk, Albert Einstein, thought this idea was worth pursuing, and his theory of relativity, where all non-accelerated frames of reference are equally valid, was the result.

Wonders in a cup of coffee. Symmetries on the surface of a coffee cup, jiggling to sound vibrations in the air

DINNER CONTINUES

Another general reflection can be provoked by consideration of the swirling cream in a stirred cup of coffee. It was first put into exact form by the Dutch mathematician, L. Brouwer, in 1911, when he proved a theorem which can be interpreted as saying that however intricately you stir your coffee, however much you twirl and twist, there will always be at least one point on the surface which is not moving. To a great extent Brouwer's theorem has been at the base of the 20th-century science of topology, the study of how certain properties will not change on highly contorted surfaces. Look close at the next cup of coffee you stir, and you can confirm the 1917 result by finding the one fixed point on the surface. If you don't find it, either you're not looking hard enough, or Brouwer's theorem is false, and the foundation of 20th-century topology, with all its applications, collapses.

At this point a demon enters the room, a very famous demon, one first conjured up by the Scottish physicist James Clerk Maxwell. Why is it that the coffee always seems to heat up the spoon? We could imagine the hot coffee just sitting there, swirling gently around the cup, and the metal spoon inside quite unaffected, staying as cool as it was when put in. All we would need would be a mischievous demon in place on the spoon surface, perhaps with crowds of subaltern demons to help, to kick away any hot molecules that looked like heating up the metal spoon before they got too close. A spoon so protected would never get hot, however long it was immersed. In real life this never happens of course, but why? What is it about the structure of the matter that makes heat always flow from the coffee to the spoon? Counter-demons perhaps?

Just about. Building on Maxwell's ruminations and other ones, 19th-century physicists came up with the notion of entropy, the idea that everything in our universe is so constructed that systems of high complexity will inevitably turn into systems of lower complexity. An arrangement whereby hot coffee formed a separate band around a cool spoon is a system of very high complexity indeed, as the recourse we must have to an active demon to keep it from breaking down makes clear: far less complex would be for the heat energy in the coffee and spoon to merge, for the coffee to lose some, the spoon to gain some, with everything in the cup ending up at the same, intermediate temperature.

If this were confined to coffee cups there would be no reason to make much of a fuss over it, but the thing about the entropy concept is that scientists hold it to apply to everything in the universe. Place any two complex systems near each other and, speaking roughly, in time they will run down, merge together, and end up as a single blurry thing, far less differentiated than what had been before. This happens to human beings — the heat from our intricately churning brain cells ending up just amorphously in the air around our head — and given the entropy concept it must happen to cities, planets, the solar system, and indeed our whole universe. The sun at the center of our solar system will in time cool down, as indeed will the other suns that make up the rest of our galaxy. In time everything in the universe will even out, settle down in an undifferentiated haze at one temperature and, from that position, be unable to change, create stars or life, ever again. This is called the ultimate heat-death of the universe, and if entropy is true, it's inevitable.

This entropy-forced running down of things comes about in a world where all matter, everything we see around us, is made of atoms which are composed of central core regions isolated by great spaces from the much smaller electrons that form the outer part of the atom. The spaces are vast indeed. Inside just one of the carbon atoms that goes to make up your body or coffee, your furniture or house, the nucleus at the center of the atom has a width that is less than one-trillionth of the distance to the outer edge of the atom, where the electrons are. If that central nucleus were enlarged till it was the size of a bowling ball, the electrons would be off somewhere dozens of miles from the center. The volume between them and the core would be empty space. The coffee spoon that seems to be so solid, so filled with stuff, is actually just a great open volume, with the occasional atom cores and far distant electron clouds spreading around them. Most of what we think of as spoon is not spoon at all but empty space within these cavernously empty atoms; most of what we think of as our bodies or homes are such empty space, marked out by these occasional atom cores too. Fingers drumming aimlessly on a coffee table are just one region of near-nothingness streaming down and vaguely interacting with another, only slightly different, region of near nothingness.

It is a somber speculation, but it has one consolation. Keep it up long enough and, wonder of wonders, when you do finally look up from the coffee cup, the odds are that the guests will no longer be there to see.

Thermogram of a bedside lamp, fading out

B A T H & B E D

SIX

After the day, after the guests, after the hosting, standing, sitting, smiling, speaking, serving, clearing, and being smoked upon, the time has come to purge it all. Clothes that had been carefully put on are now hurriedly tugged off; watches, wristlets, and possibly chest wigs are tossed beside them, and then the pink birthday-suit clad individual, pausing not even for a final temperature adjustment of the gushing taps, is ready to jump in the tub, where a luscious hot soaking bath awaits.

Is this wise? If water were just a wet sudsy fluid that sloshed about in the tub there would be no problem. But it is not. The water in your tub is not a normal fluid, not a collection of separate molecules rolling around each other as we expect our fluids to be. Rather it is a thing, one very large, quite interconnected, tub-sized chunk. Each water molecule in the tub has velcro-like electronic attachment points on the outside, one on each end, and where two water molecules are at all close to each other the velcro attachments hitch up. The connections are not entirely rigid — they're pliable, easily bent, and with a good push will even pull apart — but they make the freshly filled tub into probably the largest single 'molecule' you will ever see. The water that fell out of the faucet was stuck together in this way, and fell tubwards in what might be thought of as a continuous glob. Since your water supply probably comes from a lake or reservoir you can think of your filled bathtub as a connected pseudopodal extension of the central water 'organism', stretching its members like a mammoth amoeba through the maze of pipes to your home.

It's into this single gloop that the bather's heat-wary toe descends. If the water molecules' connections were just a few per cent stronger the water blob would be linked as tightly as iron, and the toe would find itself resting on a shiny – and iron-hard – metal surface. If the two per cent plus transformation took place after the toe were in place, then a blow torch would be called for to get it out. As it is, the bonds are pliable, and as the toe descends it stretches the nearest ones, lengthening them like bread dough being pulled sideways, until finally the press is too much, and with the sub-microscopic equivalent of a rip the continuous hunk of the bath water thing suffers a single slit into which the wriggly-toed foot protrudes. As it descends it continues to stretch and rip the water agglomeration in its way, but because of the water's rebounding snapback, it quickly closes behind the toe, wrapping it, and then the foot and soon the collapsing whole body in a snug hold. There the entrapped bather can lift his leg, splash his hands, play with a duckie or even scramble for that soap which has the habit of uncomfortably descending to obscure places. All give the impression that no force at all is wrapped tight around and resisting him, but in fact all these movements are due not to the good-natured grace of the enveloping water thing, but merely to the fact that the bather is strong enough to continually slit open and drive through those gaps in the otherwise monolithic water solid when he wants to move. If he were ever to become too weak, there would be no escape.

Even if the occupant escapes, the bath oil on the surface might not be so lucky. If it's a thick and heavy oil, it will drive through the weak points in the water, but it will then also soak the person luxuriating in the tub in a heavy layer of grease. But if it's a lighter and more gentle oil, one that won't make the bather feel as if he has entered a tanker spill area, then the water has a chance to get to work on it. The water resists the light oil's advance, changing it from a thin film to individual lens-like globules. This is not an improvement, which is why satisfactory bath oils are hard to formulate.

Swooshing through the water in addition is the sponge. Some bathers have those cubical factory made ones, but true connoisseurs insist on natural. Most sponges we handle, fingers poking firmly in the holes, were originally female when they were alive, though there are also many bisexual ones around for sale. The holes, though, are not what one might think. Sponges have them because they lack legs, and so being immobile can't clamber over to any free-floating food in their sea-bottom habitat. Rather they have to pull it in. The major holes are to allow water to enter, and living inside them are great populations of propeller-kicking flagellar cells, stuck like latter day galley slaves, their whole lives spent rowing the water in. Hiding deeper inside the holes is another population consisting of free-moving amoeba-like cells that pounce on any food the outer oar-beaters have managed to bring in.

That set-up alone would produce a pretty miserable-looking sponge. A rubber ball with pock marks on the surface for flagellators and pouncers would do. The reason sponges that end up in our bath don't look like rubber balls with acne is that if all they did in the wild was pull in water, they would soon swell up, resembling balloons and in time even undersea dirigibles before they burst. Sponges avoid this by having as many exit holes as entrance ones, to get the excess water out, and lots of internal linking tunnels to direct the incoming water to those exits. And, for a sponge, that's it. There's no brain, liver, shoulder bones, muscles, teeth, jaw, smile, eyes or anything else. Occasionally there's a gushing out of bobbing eggs, and for those species so equipped something like distorted sperm cells too. But such reproductive outpourings are rare. What we have is a creature good only for soaking up water and pushing it out again – exactly what we want once the living cells are cleaned away for a handy water moving tool in the tub.

It seems curious, and so it is. Very few people willingly get into the tubs where this water thing, oil and sponge are to be found. A survey of 2,000 British people showed that men hardly ever bathe, needing extreme compulsion, such as a breakdown in the shower's working, or a trying evening with a recalcitrant dinner and food-slobbering guests, to enter a tub. Children both male and female are equally reticent. The only categories that seem to plunge in without duress are teenage girls, and the most frequent of all, adult women.

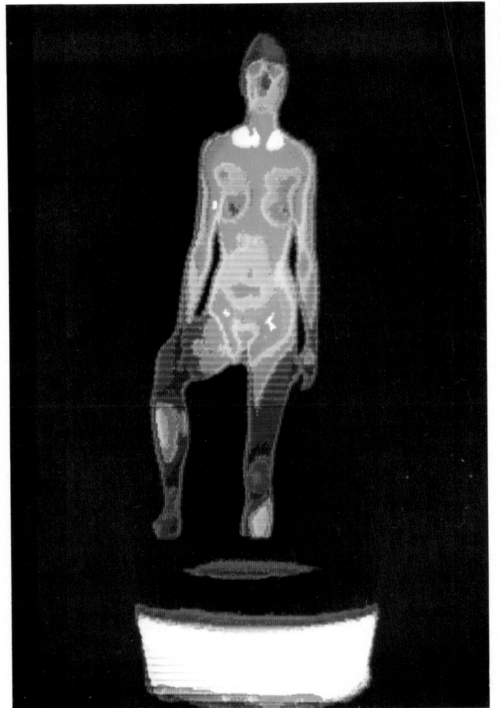

Clockwise from left: Heat images of a woman entering a bath, soaking, stepping out and drying. White and red are hottest, green and blue coolest

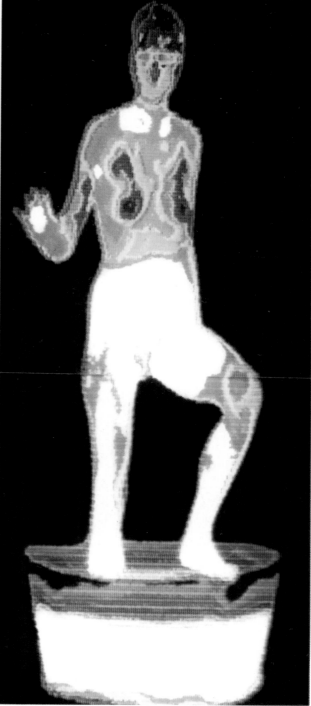

The women who do bathe more do so primarily at night (80 per cent then, as versus 20 per cent early in the morning), though the timing habits of the first noted English bather, Elizabeth I, are not recorded, except that she resolutely took one bath a month 'whether she need it or no'. After her time there was a long hiatus in popular western bathing. Hollywood presents a different picture, with courtly ladies stripping off and jumping into the nearest stream whenever they could, but this was largely a way of introducing some corporeal interest into film despite the strictures of the censoring Hay's Office. It has little connection to fact. Northern rivers are cold. Women at the French court did apparently try to bathe in the Seine at its nearest point to Versailles once, but it was not a great success.

The modern bathtub was apparently developed as a punishment device for lunatic asylums early in the 19th century. Only after the spread of easy water heating later in the century did it move to the home. Bathers of course had little idea what to do with these curious items when they were reintroduced. There were instruction manuals: how to step in, how to sit so as 'to avoid undue shock to the genitalia', etc. Showers were approached equally warily. In the 1870s in Britain and America iron 'bathing helmets' were recommended as protection from the onslaught. Even great men trembled. Disraeli's wife revealed that her husband was so scared of the cold jet to come that he would stand in their home shower booth cowering in the nude, incapable of turning on the flow, until his wife reached in a hand and pulled the nozzle chain to start it for him. She herself, sensible woman, took baths.

The pleasures of a bubbly soaking, of lying back in luxuriant warm peace after the guests have gone, are, in time, followed by the fascination of watching the whirlpool of exiting water spiral down the drain; a fascination aided by the numerous questions it suggests to the inquisitive now clean and clambering outward occupant. Why does the whirlpool make that funny gurgling noise? How come it forms again after you break it by poking your finger in and wiggling it around? Is it the same as the whirlpools that form in the oceans? And, above all, is it really true that a bathtub whirlpool turns one way in the northern hemisphere, and the other way in the southern?

To get an answer we need to look at the Coriolis force, named after the French scientist Gaspard de Coriolis. This is an account of the slippage all moving objects suffer due to the fact that the earth is spinning as it moves in its orbit. At the equator the earth is spinning to the east at 1,041 miles per hour, and as you go north that rate diminishes. In Florida it's spinning at 920 miles per hour, in London or Warsaw (on the same latitude) it's 620 miles per hour; only up at the North Pole is the turning rate zero — for there is the axis around which everything else is turning.

This means that as you sit musing in your bath in London you travel over 900 feet to the east every second; on vacation in Portugal, being further south, the same horizontal musing would put you an even greater 1,200 feet to the east. If your momentum were independent of the earth, the consequence would be that you would hurtle at high velocity through the wall of your house or hotel and out to the east at this terrific rate, clad only in a smile and some residual bubble-bath foam. As your momentum is not usually independent of the earth, this embarrassing event is accordingly unlikely to happen. Only if you were not a pink-skinned bather, but a free-moving artillery shell, would matters be different.

Fire a giant canon in Florida that's aimed precisely at New York and, because of the faster horizontal speed the shell has at its start down south in Florida, it would miss New York and land many miles to the east in the Atlantic Ocean. (This extra horizontal speed the further south you go is why NASA chose to put their launch site in Florida — the furthest south you can go in the US — just as Jules Verne predicted in his fantasy *From the Earth to the Moon* in the last century.) For a ship firing a shell just 15 miles the slippage from the earth's rotation is not so great, perhaps 100 yards, but that's still enough to convert a clear hit into a useless splash in the water unless it's accounted for. Since Britain in the First World War had the greatest navy, technicians in the Admiralty design department were aware of this effect and had taken it into account in their warship blueprints by having the shipboard gun turrets automatically aim to the west in firing by just the needed amount to compensate for this Coriolis force. At the 1914 battle with light German forces off the Falkland Islands, down in the southern hemisphere, the guns were given the appropriate reverse compensation, and the Germans were sunk.

This effect applies to fluids as well as to 2-ton shells. Our mythical shell released northward from Florida would veer to the east before hitting New York, and so would a blast of water sent gushing northward on the surface of the Atlantic from Florida. The only difference is that this water flow really exists: it is the Gulf Stream. Starting out due north from near Florida it gets twisted over to the east as it moves, entirely because of the Coriolis force. As a result England is warmed; if the earth spun in reverse direction the stream would start off the coast of Ghana in Africa and end up in Labrador, so warming that province of Canada and leaving England to freeze. Tornadoes and hurricanes rotate clockwise in the northern hemisphere and counterclockwise in the south for the same reason.

Is this what accounts for the direction of water spinning down your bathtub drain too? Is it really one way here, and the other in the opposite hemisphere? Unfortunately the Coriolis force does not often get to show up in the tub: it's usually too delicate on that restricted scale. Any little splashes you make in pulling yourself up, or even the disturbance your now prune-puckered foot makes as it gropes for a steady placement on the bottom of the bath, will be enough to start the drain whirlpool spinning in the direction of that splash, and over-ride the Coriolis force. Only if you got up very, very carefully, without disturbing the water more on one side than the other, would this force that counteracts dreadnought shells and massive ocean currents get a chance to force your bathtub's moving water into a pure Coriolis *clockwise* gurgling whirlpool. Watching it you might well marvel that a practitioner of the same experiment in Australia would be seeing his carefully emerged-from bathwater gurgle down in the opposite direction.

If that delicate extrication proves impossible you can still contemplate the fact that as you slosh and stumble all wet, oil-coated and warm from the tub, the Coriolis force is pushing you sideways to the right with every step you take. For a two-foot stride it comes out to a few thousandths of an inch diversion to the side; in a 26-mile marathon it would be several inches; and in the estimated 3,000 miles you might walk in a decade there will be several hundred yards of these Coriolis-forced staggerings to the right in all. For car journeys it's just the same: put 200,000 miles on your car and you will have without notice slid quite a few miles to the right, and just as much without notice corrected the slide each time before it could build up.

While the man is climbing out, bath over, body scrubbed, the woman is contorting her brow and glaring into a mirror, cotton pads, ablutives,

A soap bubble, one of the thinnest substances the eye can detect: two layers of fat with water trickling between them. Schlieren photography enhances the iridescent colors produced

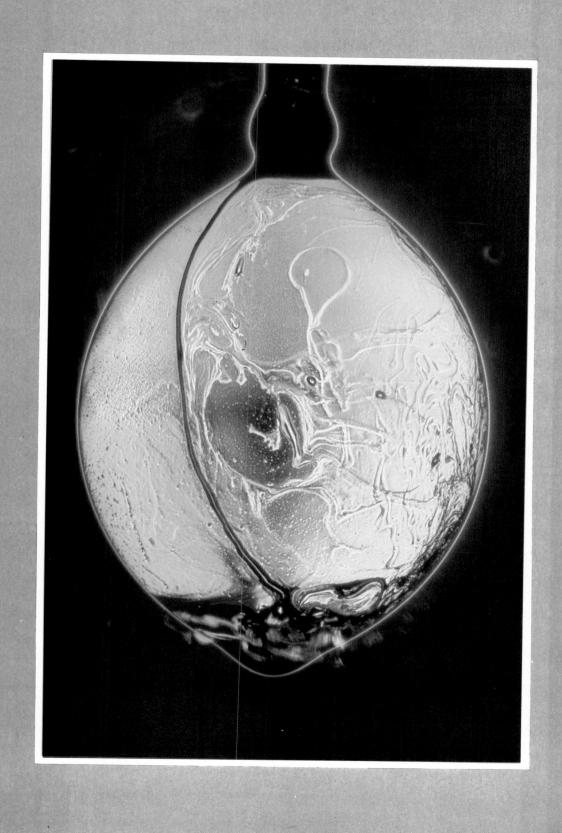

and other stripping chemicals at the ready, trying to do something about her face. This is not easy. There can be much curious covering on the conscientiously prepared hostess's face — rouge or face powder, mascara, cake make-up, foundation cream and maybe more; layers upon layers, sheets of oil over firmaments of colored talc, veritable geological strata written on the human form. Even that is only a start, as under these adornments are the materials a face itself is likely to produce by the end of a long day, items easily seen through a microscope or just a good magnifying glass: the skin cells and flakes, excretory pore soils, pus, blood cells, serous exudates, and occasional crust to fill the spaces in between. And, outnumbering all else, there are the face bacteria.

These have been on your face all day, many millions of live bacteria creatures, another individual doppelganger, stirring and restless and there for inspection by any colleague carrying a microscope. Some hang from the face on swaying strands, others roll and bounce across the surface, while yet others, like the morning pseudomonads, are cranking their propellers for all they're worth, trying to migrate a vast distance, perhaps an inch or more, from one region on the face to another. The individuals are only a few hours old, though the populations have been there since your birth, and sometimes before — types of bacteria specific to your mother's birth canal are near certain to be around. They will be elsewhere on the body too — perhaps 8,000 per square inch on the legs, 40,000 per square inch across the chest — but nowhere near as many as on the face, where typical day's end counts are over 2,000,000 distinct creatures on your cheek, chin and nose, and more on the forehead. They prosper when we do, freeze when we do, and many will even die when we do.

Faces are such attractive places for bacteria because the conditions there are just what these creatures need. The individual skin cells that make up the outside of our face are about 14 microns ($^{14}/_{1,000}$ of a millimeter) in diameter. (Bacteria are often 1 or 2 microns long.) These cells are hardened and flat by the time they reach the surface, and are indeed what rubbed off in such great quantities to float around the room when we dressed, as seen in chapter one. Now if these hardened skin cells covered the face without a gap there would be no way for the bacteria to live there. They would starve. But your skin cells are so far gone by the

time they reach the surface, so ragged and fraying and half broken off, that there are lots of gaps to the deeper layers below, and it is through those gaps that food the bacteria need pops up. Combine that with the tropical 90 per cent humidity immediately surrounding the face and it is ideal.

Imagine how easy life would be if all you had to do was loll on the floor of your house, and neatly on a regular schedule sandwiches would pop up from cracks in the floorboards to supply your every culinary want. That's the luxury the bacteria on your face get to experience. The food welling up from cracks in the skin is not our usual lunchtime fare, not ham and cheese or pastrami on rye – those are substances the bacteria here have to miss – but rather more suitable stuff; some salt water from the sweat pores, nitrogen compounds conveniently mixed in with it like a vitamin seltzer, and wherever there's a hair follicle some oily face grease will be plopping up too, grease just loaded with the amino acids bacteria need.

Because it's all on such a minute scale the substances don't act quite as we're used to – thus the apparently deep salt water pools can suddenly evaporate in just a few seconds, so leaving any bacteria that were too slow to drink their fill dying of thirst as this real mirage goes – and of course the bacteria don't act as we expect either. To see what's going on imagine a man stretched to 200 miles tall, like Voltaire's Micromegas, or the Jolly Green Giant with thyroid problems, his head sticking out of the atmosphere, at the level of the Space Shuttle's orbit, his feet standing 100 miles apart. On that scale the bacteria on his body and face would be ten inches long, the size approximately of newborn Dobermann puppies; the skin flakes they're resting on the size of a 12-foot concrete yard.

For a few minutes those Dobermann bacteria would be fairly normal, lapping up salt water from the encrusted sweat pore tunnels opening onto the skin surface, possibly rolling and frolicking in the sticky sebum on the surface too. But then, after about ten minutes of this, one Dobermann would stand up with a curious anxious look in its eyes. It would begin to breath in, narrowing and narrowing around the waist until suddenly, as you were looking, it split in two, the front half growing legs to become a fully-fledged Dobermann puppy, and the back half growing a head to do the same. For a ten-minute lifespan this splitting would take about 30 seconds, and without delay the two new Dobermanns would go back to drinking and frolicking as their now vanished parent had been doing. The same happens to all the other bacteria, and it's because of this rapid growth rate that they can survive the slaughter that occurs when you aimlessly touch a finger to your head – the equivalent of a jetliner plopping out of the sky to land on these Dobermanns – or, worse, when you commence the country-wide devastation of actually slapping a hand to your forehead to look closer in the mirror.

There also is the problem that the whole setting on the face is unstable. Unlike what we would expect if lolling about on a house's floorboards, the dried skin flakes that make up the surface of the face regularly break loose and fly off into the air. It would be as if we were resting with some friends on the large concrete blocks that make up the surface of a public park, and every now and then some of these blocks just lifted up and soared away, carrying our friends with them. Such skin eruptions are happening all the time. Back at the dinner table, a good number of your bacteria-laden facial skin flakes will have lifted off and landed on the guests – as indeed a number of theirs will have landed on you.

Some face bacteria try to avoid this problem by burrowing down deep, under the so unstable surface skin flakes. These are especially interesting because they are direct descendants of the very first bacteria that evolved: creatures from the era when oxygen was still a rare and universally poisonous substance. Since your forehead residents have not evolved beyond that point they have to hide from oxygen even today, which is another reason they prefer the forehead. There is a persistent layer of surface grease there – thin to us, but a thick rubber webbing overhead to them – and this sealant is excellent for keeping the unwanted oxygen out. It's safe under that greasy seal too, as most other bacteria would suffocate without air, and it's for this reason that the forehead dwellers are among the most prevalent bacterial creatures on your face, reaching densities of about eight million on each square inch. On the nine or so square inches of normal forehead there will be 72 million of these ancient oxygen-fearing dwellers in place. They clump together in little shaking heaps of several thousand, but even in those agglomerations are still just too small to make out with the unaided eye. Wear lots of make-up and these oxygen-avoiding bacteria will have even more space to thrive under their protective grease layers.

If all this information makes you feel uncomfortable, a frantic effort now to scrub these creatures off before anyone else could see you would not be worth your while. The bacteria are perfectly harmless where they are, and on the model of the new arrivals in a devastated armpit it could even be dangerous to get rid of them entirely. Some would be killed, but many, hiding just below the surface in the pores from where their food comes, would just wait, however hard you scrubbed, until the fury was over, and then they would emerge again to recolonize the so kindly cleared space. Bacteria numbers often go up after a wash, as epidemiologists studying surgeons' about-to-be gloved hands have found to their dismay.

Far better is to take comfort knowing how petrified everyone would be if they were to get a magnified glimpse of yet one more of the creatures that live on your face. These are giants compared to the others, not simple bacteria but the massive armored mites, creatures 30, 40 or even 50 microns across. They are hulks that permanently live in your eyelash bases, grasping on to the hair there with their eight gnarled and strong legs. Living in such an odd place they were only properly divided into species in 1972, but studies since have shown them to be present on virtually all of us. With nutritious eyeliner applied they flourish even more. They're hanging in place as you go to any meeting, ponderously swinging around in whatever direction you look. Cautious hunters, they hardly ever crawl all the way out of the eyelash base — except to migrate across our peacefully closed eyelids later at night, the better to find another follicle home to move into, and perhaps mate. During the day, however threateningly your fellows glare, they won't budge.

Although we're stuck with the bacteria, the make-up and other gunk on the face can be removed. Soap and water won't work to get it off, as the grease from the make-up gets in the way. Something is needed to work through the make-up, to pull it away and reveal the surface underneath; something the Roman matriarchs, who went to bed stoutly defended in briefs (*strophium*), bra (*mamillare*), corsets (*capitium*) and tunic, called their *ceratum refrigerans*, and which we, usually less encumbered, know as cold cream, or cleansing cream.

Central to the cleansing cream the woman now applies is mineral oil, 50 per cent in most mixtures. This is a relatively light paraffin residue – it

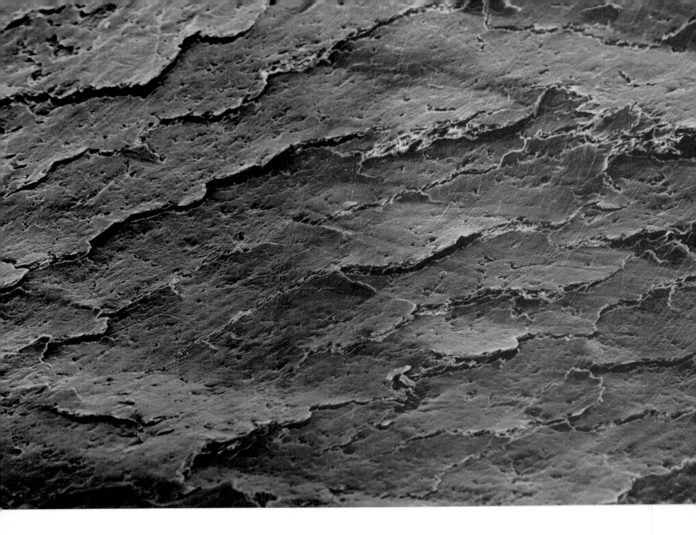

floats on water — and rubbed on the face it slips around any make-up grease and lifts it into suspension. If that were all there were to cleansing cream a certain customer resistance might be encountered. Mineral oil is so light that any grease it eases up is likely to soon slosh back down. The resultant goo would be like having Brylcreem spread on the face. Something is needed to hold the make-up grease in suspension until it can be wiped off, and that is what provides the second ingredient in cleansing cream: beeswax. This is the regurgitated stomach contents of female worker bees, used, in the hive, to provide the structure where mucus-coated larvae can rest. In the cream it wraps around and holds the mineral oil plus attached gunk, producing a stable margarine-like emulsion for however long it takes an absorbent tissue to be picked up, then wadded onto the face to wipe away the mess.

Surface of a bar of soap, ready to flake off on rubbing

To complete the cleansing cream there is lanolin – the polite word for sheep sweat. It's not there for the odor (as proximal investigations of shepherds on duty will make clear) but rather because the lanolin molecule is blessed with almost the same mixture of free water-binding sections as there are on the back of a postage stamp. Supply some water from the tongue to a stamp and these water-binding sections will generate an adhesive film to stick to any envelope they're pushed down upon; supply some moisture as might be left over from the kleenex's passage to the lanolin, and its water-binding groups will create a see-through plasticy wrapping, firmly enough attached to keep the cleaned one covered through the night. This, for some, is a treat.

When the man finally gets a chance to use the sink area, leaving his domain of strewn towels and stranded plastic ducks behind, and replacing it with the realm of open cosmetic jars and wadded cotton puffs, his ablutions are likely to be less dramatic. A wary splash of some cold water is most likely; possibly an admixture of soap too if he is brave. This is understandable, a response to all those enforced scrubbings behind the ear in youth, but still a mistake. Water alone does not do a great deal. It's not that tap water is repelled by skin grease – the sebum our face oozes out has loose molecular fronds that are keen to bind with water – but rather that water takes such a long time to get in all the little folds in the skin where dirt is especially likely to have been stuck. Even when it does reach the dirt there will be little actual dissolving in the water: all that happens is that the dirt is splashed from one niche on the face to another, redistributed rather than removed. Bacteria are equally resistant to being dissolved, and so also just get tossed about on the face when the water is heaved on.

Soap is better. It has micelles. These are little doughnut-shaped formations of molecules that stream off the soap bar in terrific quantities, many, many trillions with every rub. At first they just flop through the water in the washer's hands, spinning futilely, but when that handful of water is slopped against the face, they get to work. Landing on a portion of the face where there's some dirt they envelop it tight and don't let go. Let the water slop back into the sink, and the grime-embracing micelle doughnuts will go off with it. More soap, more cleaning, and after a few bouts of sudsing the facial terrain, so recently coated with those skin cells and flakes, excretory pore oils, pus etc., is much clearer than before.

It's a neat trick, and works with armpits, ears, toes, textiles and much else too. The soap that could do these cleaning miracles has accordingly long been prized, which was a problem, as it was very difficult to make. To get soap you need something gooey, such as boiled fats, combined with something that will turn that goo into the effective lifting micelles. Boiled fats are easy to get; it was the other something that provided the problem. For many centuries the ancient Egyptians used impure soda deposits from the great Wadi Natrum in the Western Desert. The soda they got there (called natron) was able to combine with fats to make soap, but it was a miserable sort of soap, as a natron and fat mixture only produces fragile micelles — weak, delicate things that peter out after lifting up a small amount of dirt. Still it was the best anyone could do, and when not used in embalming (Tutankhamen and others were embalmed with natron soap) became an export item to Europe.

There was probably enough natron in the Wadi to keep everyone in Europe clean for a long time, but the Ptolemaic authorities knew a good thing when they saw one, and put a high tax on natron exports, so pricing it into the luxury market. Barbarian courts went wild with the stuff, using the foreign soap yearly, twice yearly, and sometimes, audaciously, if decadence had set in and the female influence overwhelmed, even once a month. For the 99.99 per cent of the population which couldn't afford the natron imports though, ineffective cold water splashings were made to do, or perhaps a type of local soap made from collecting ashes. The ashes being habitually burned and collected in large pots, the result was called potash (*pot ash*) soap. It is worse than the natron stuff, having micelles that fall apart upon contacting dirt. Only since no one bathed much, was there little problem with soap ingredient supplies being so hard to get. Occasionally there were gluts in the soap market — there are records of 1,500 camels being used to haul natron from a freshly discovered deposit of it near the river Hermus to soap factories in Smyrna in the 17th century — but usually supplies stayed low, and no one saw ground for complaint.

In the late 18th century that complacency towards soap ingredient shortfalls changed. This was not because of some sudden fetish for personal hygiene, some shame at encrusted dirt and desire to be pink and glistening, but rather because the soda that both natron and potash are composed of began to be found useful for glass-making, for textile preparation, and, above all, for high explosives. Such a potential lack in military supplies was serious, and all the great powers began desperately looking for alternate supplies of the stuff beyond what tree ash or the Egyptian Wadi could provide. The soda panic was perhaps equivalent to what would happen today if militaristic governments found they were about to run out of uranium supplies. Researchers found that kelp supplied an especially rich mixture of soda when burned; accordingly claims began to be staked on islands where kelp strands were especially abundant, and it was to some extent for this reason that Britain saw fit in the early 1770s to threaten Spain with war over the Falkland Islands, which to many at the time may have seemed some desolate mid-ocean rock outcroppings, but which to the strategic mind became crucial because of their abundant supplies of nice, festering, soda-rich kelp.

Matters got so rough that in 1775 the French government announced a fortune of an award to the first person who could come up with a way of making soda in a factory, so ending this hunt among the wadis, forests, and kelp beds which the English navy were so good at protecting. Now there are awards and there are awards. Let a small manufacturer announce an award and you can't be sure he will pay up in the end. But when it was the French monarchy announcing a prize — this monarchy which had ruled in unbroken succession for almost a thousand years — then you could be sure that however long it took to come up with an answer, however many blind alleys you had to explore, however much of your personal savings you had to put in before the successful process was achieved, they would be there in the end to accept the solution, bestow the honors, perhaps bring on a feast, and above all pay you what was your due.

That, at least, was the woefully mistaken reasoning of one young surgeon from the French provinces named Nicolas Leblanc. As soon as the award was announced he started looking into the problem of making soda. He went through alchemists' old studies, he visited soap factories, he ordered quantities of the Egyptian natron; he did everything possible to secure the prize, and when his experiments on reproducing it in a laboratory came to nothing in the first year, he didn't give up but kept on going, for another year, and then another and another, certain that in the end he would crack it. And so he did. After 12 years of effort, in 1787, Leblanc perfected a process that could produce artificial soda on a commercial scale. Being eager to protect his effort he got a royal protector – the Duc d'Orleans, whom he had worked for as a physician – to ensure that there was no hanky-panky at court that might keep him from receiving his just reward. Leblanc built a working model to demonstrate his process, officials of the King's Academy of Science came over to see that it worked, the proper bribes were discreetly distributed, more officials came to see, and then, just as he was about to be granted the award, just as all his work was about to pay off, in the summer of 1789, some fools went ahead and made a revolution.

The Science Academy officials were sorry, they personally would like to help, but until they knew what their own position was there was nothing they could do, Leblanc must understand. Everything suggested it would be but a brief delay. Already there were sensible voices in Paris suggesting moderation, leniency to the King, a reprimand perhaps, but otherwise no hard feelings and a swift return to business as before. This party had Leblanc's fervent support, and there's some evidence it might have prevailed. But . . . the Duc d'Orleans, Leblanc's own chosen protector, thought otherwise: if there was going to be a revolution he wanted to be at the head of it. The King had been captured; he ought to be sent to the guillotine, the Duc cried. He was. That meant another delay in paying Leblanc. Then the Duc himself was sent to the guillotine; still more delays.

It was at this point that Leblanc's luck took a decided turn for the worse. Why not, he had thought at the start of the revolution, trust in the good intentions of the new government, and start up a factory to make the soda even without securing the reward first? Whatever the outcome it couldn't be worse than what had come already. Just to protect himself Leblanc applied for a new document the revolutionary government had set up to help protect his interests, something called a 'patent', and then proceeded to build, in the outskirts of Paris, a full-scale soda factory. Alas, his chemistry had been perfect, and the factory worked without a flaw. It was, accordingly, appropriated by the government without compensation when the Duc was killed. Then for some reason it closed.

This was not the wisest of acts – the soda lack got so bad by the following year, 1794, that the Committee was asking for the burning of all useless vegetation in France to get more potash – but the era was not one in which cool heads prevailed. Leblanc was now ruined, without anyone to turn to. His luck was so bad that even an English spy who came over to Paris to get news on the new soda process managed to bribe the wrong Frenchman – a competitor of Leblanc named de Morveau, who was working on a different, and inferior, process – so Leblanc did not even get the money he could have earned by treachery. In 1800 Leblanc briefly got his factory back, but by then, according to a document of the time, it was overgrown with weeds, and he couldn't raise the money to start it up again.

In later years the artificial soda process Leblanc had come up with went on to become the foundation of Europe's heavy chemical industry in the 19th century, and, through its effect on improving hygiene, helped end the epidemics of typhus, tuberculosis, and other infectious disease which had been the scourge of mankind for ages. Leblanc himself, broke and without hope of seeing the revolution overthrown, committed suicide. Even in this the poor man miscalculated. His decision to die came in 1806 – just nine years before the Battle of Waterloo and the restoration of the monarchy which, with attached Academy of Science, was quite willing to pay the original prize to the noble benefactor who had invented the artificial soda process for France . . . if only he could be found.

Retiring to the bedroom after pre-bed cleansing seems the most natural thing possible, but until quite recently it would have been out of the question. There were no bedrooms to retire to (they hadn't been invented yet), and anyway there was the question of what the guests would do. Dinner guests in mediaeval and Renaissance times couldn't be expected to head home after dessert: the roads were far too unsafe, as we've seen. Muggers, though lacking Saturday Night specials, made do quite well in those days of no police or street lights with stillettos, swords, garottes, very long pieces of wood, and the odd metal-spiked mace. Guests would expect to sleep over till dawn (as embassy personnel in dangerous countries where there's a curfew expect to even today), and, there being no other space available, would just stretch out in the feast room where they ate. Their hosts would stretch out there too.

This might seem coarse, but there was etiquette. A generous host would let his guest sleep on the bench, or maybe even on the table. Only a selfish, uncaring one would force the guests to sleep on the floor, where the servants were, down there with the remains of dinner, the rodents, moisture and drafts. It is from these huddling times that our term of 'making a bed' arises. Straw was handed out for the sleepers to spread beneath them, or, in the case of nobility, to stuff into sacks. Upon these they lay, and the bed was made. (We moderns, who according to US Department of Agriculture figures spend 25 hours and walk four miles on average per year to make our beds, should perhaps pause before scoffing.)

Even once the guests were safely locked away in separate bedrooms, the best arrangements for tranquil sleeping were still hard to work out. Servants, for example, would often sleep in their master's bedroom, which could present problems. Pepys in the 1660s records his wife repeatedly complaining that he refused to close his eyes until the young maids who slept in their room had finished undressing. In France the servant would sometimes be pressed away in a trundle bed that collapsed and slid under the main bed. The question of single versus double beds for couples did not arise, as the waste of building two individual beds where one large one would do was never considered. (Only back in Roman times, with their little night-time *cubiculum*, had single beds been in fashion, and only when American manufacturers realized there could be double profits in convincing couples to sleep in separate beds did they see a popular resurrection.)

All this sharing of bed and bedroom led to a certain easing of manners. There's the account in Aubrey's *Brief Lives* where Sir Thomas More invites Sir William Roper into the family bedroom to choose a wife. It was early in the morning, and More's two teenaged daughters were still asleep. More removed the sheets over the two girls, which revealed 'them with their smocks up as high as their armpits'. They slowly turned over, not bothering to pull down their smocks; Roper said 'I have seen both sides', patted one on the bottom, and said she would be his wife. Practiced today both More and Roper would be put away; then it was considered quite natural. Night clothes as a garment against the peering world were only

Bedbug, in full hunting glory: surface hairs for sensors, long proboscis for piercing skin. The head is tiny, but the body massive – and expandable – for holding the blood it siphons in. Endemic in every country except for a favored few

felt necessary in the last three centuries or so. Our more recent removal of them, helped by central heating (and perhaps the model of Marilyn Monroe, whose press agent induced her to say she slept in Chanel No. 5 and nothing else) is only a reversion to form.

With an individual bedroom, double bed and removal of most if not all garments, we're just about ready to get between the sheets. On cool nights a hot water bottle might be needed for those of a delicate constitution, and in the poorly heated homes of Britain this device became near universal, prompting the Hungarian-born satirist George Mikes to note that where the Continentals had sex, the English had hot water bottles. Gladstone sensibly used to fill his bottle with sugared tea, so providing a handy source of sustenance on the morrow as well.

The water bottles below might be matched by 'paintings of a licentious nature' on the canopy above, a practice common at the French court to '*encourager les plus refroidis*'. The geometrical suggestion was that it was the man who looked up for aid. For a time under Louis XIV these helpful paintings were replaced by the even more live action mirrors. Unfortunately this advance was too early, and as recorded by the industrious Lawrence Wright, from whose work much of this information about beds is drawn, one Monsieur de Calonne was so rumbustious in his bedtime activities that the vibrations caused the mirror on his canopy at Versailles to fall, almost vivisecting him. The court went back to still-lifes, and the presence of mirrors, for those so inclined, had to await the advent of wall or cupboard mountings.

Bedroom dressing tables where watches and other valuables could be put away are also fairly recent, and had to await the invention of the sliding drawer. Before that, daily accumulations were always put in padlocked boxes under the bed. Mediaeval registers show coins, clothes, hats, frying pans, saddles – the equivalent of car keys – boots, books and other valuables stored in the bed box.

Finally our couple are ready for bed. But the woman is not ready for sleep. A certain sly, lip-full smile suggests what is to come. The woman extends one bare shoulder out from the covers, longingly reaches over . . . and collects the tempestuous romance lying on the night-table to read.

Centuries ago the pages of such a book would have been made of parchment, a substance that had to be shaved of hairy stubble before it could be used, made as it was from the skins of goats, lambs, and other hirsute beasts. Only with the development of paper-making did smoother sheets become the rule.

Yet all is not well on these hairless sheets of contemporary paper. A curious projectile loose in the bedroom is screaming down from the air and exploding on the book's surface. The projectile is destroyed in the assault, dangling uselessly from the paper where it hits, but as soon as it is used up another swoops down and crunches somewhere else on the paper. No gentle ping-ping, it's more a frenzied rat-a-tat-tat, with fleets of these kamikaze projectiles targeting on the paper page.

The name of these miniature assault projectiles is oxygen. Your bedroom is stuffed with them, crammed and loaded with many quintillion of these mobile charges. Each one hitting produces a flash and pinprick of heat, and with the great numbers storming out of the room air each second the paper suffers the equivalent of being caught in a slowly burning fire. This is no figure of speech. Drop a sheet of paper on a log fire, as heroines in tempestuous romances are wont to do with their plot-entangling love letters, and if you look close you'll see that it turns yellow right before it curls up and actually ignites. That yellow shift is due to the assault of aerial oxygen atoms on the periphery of the fire, their action speeded by the so-close heat. With the book open in bed, the effect is the same, just slower.

This slow burning holds for every kind of paper, and is why cheap paperbacks quickly go yellow in the air. Fancier books have a thin clay spray on their pages, covering them like a sheer plastic wrap covers a stored fruit salad, and so blocking the oxygen assault. Authors prefer this, as a book with such coated paper can last 40 or more years to a paperback's 10; publishers, perhaps prompted by their accountants, take a more philosophical line on the inevitable perishability of human production, and grind out the uncoated stuff anyway. (Higher acid levels in cheaper paper makes the oxygen burning even worse.)

It's not old, this oxygen searing away in your bedroom, most having floated out of the nearest ocean, or plopped down from a convenient plant or out of the soil no more than a few thousand years ago. Fresh air is fresh indeed. When the earth was young, there was almost no oxygen around in the air, and land animals, which depend on it, could not exist. Some bubbled up from great liquid deposits in the gaps from rocks way below the surface; most started churning out a mere 400 million years ago when plants began to grow on the land and let out these reactive oxygen projectiles as a dangerous waste. The way was clear for land creatures that could scavenge and even live on this unwanted leftover: first adventuresome exploring beings like the centipede and early spiders, then their less enterprising descendants, such as us, who took it as normal to breathe the gaseous ooze which plants so carefully let away.

All together about 20 per cent of the air is oxygen now, say the equivalent of 200 gallons – four bathtubs full – invisibly sloshing around in your bedroom. If there were much more, rust would grow like dripping wall paint (rust is merely the process of iron combining with oxygen to return to the mixed state it was in before smelting), paperbacks would crumble after just weeks of handling, and a casual match struck to light a cigarette would flare like a blowtorch. Even dialing the phone to get help would be tricky, as the sparks from that could set off a blaze. If the whole earth were covered with more oxygen, any forest fire that started would spread over big sections of continents, and a fire anywhere in a city would likely spread faster than any the fire department could contain, especially as the minute sparks from a speeding fire engine's axles would probably set off their own flames wherever they landed. The effect would be like living under a worldwide oxygen tent – and indeed the regulations on use of hospital oxygen tents are harsh for this reason.

How then did it turn out that we're lucky enough to live on a planet with only 20 per cent free oxygen and no more; where we're not scared to light a match, where fire engines do not have to move at a cautious crawl, and there are no giant continent-covering forest fires? The answer is that such blazes did take place – whenever the oxygen level rose too high. But the result of those fires was fewer plants and trees – enough, with other factors, to help lower the oxygen production level to the stable 20 per cent. The charred lands back then are what produce our safe four bathtubs-full per bedroom now.

The discovery of oxygen was a long time in coming, as researchers from Aristotle on tended to think of the air around them as either an invisible nothing, or just one big chaotic gas (the word 'gas' comes from the Greek *chaos* for this reason). The realization that oxygen was a separate element, and one that makes things burn, occurred to our Yorkshireman Joseph Priestly again in the fertile 1770s. Unfortunately Priestly was a kind-hearted man, willing to talk about his discoveries with anyone who asked, and that led to an important murder. On a visit to Paris shortly after it, Priestly had no qualms about passing on news of his discovery to a certain French researcher, Antoine-Laurent Lavoisier. After Priestly had left, it seems that Lavoisier double-crossed him, and published the discovery of oxygen under his own name, or at least took more credit in it than he should have. This was not wise. Through his publication on oxygen (and other legitimate researches too) Lavoisier became recognized as the leading scientist in France, and was made director of the Academy of

Sciences. One of his tasks there was to decide which young researchers could attend lectures, and one young man whom he turned away, terming him a crackpot, was Jean Paul Marat. Several years later, during the Revolution, Lavoisier was brought before the Convention on a charge of stealing tax funds when he had been working for the King. The charge was accurate enough, but as judges sometimes took a sympathetic view of this offense there was some chance he would be let free. But this time the judges, led by a now older Jean-Paul Marat (who had decided on an alternate career in politics after being rejected in science), did not take a sympathetic view, and Lavoisier was guillotined that day.

The number of oxygen atoms in the air is surprising. There will be over 300,000,000,000,000,000,000,000,000 soaring loose in your bedroom as you turn a book's pages, each one vibrating and tumbling along in the air. They don't just attack the paper and other substances, but also clang into each other, bumbling into head-on collisions an average of once every seven billionths of a second. Being so small that the bedroom door doesn't stop them, or even for that matter the hall, the closed window, or the front door, the oxygen atoms in your room are quick to bumble outwards exiting the house and floating along above the steet. At the same time oxygen atoms from outside are taking the reverse path and filtering into your building, ending up ready to attack your paper or even be breathed after a journey that saw them perhaps 50 miles away at the start of the day. In two weeks' time the oxygen atoms that had been in your room will have mingled with the rest of the atmosphere and be over 1,000 miles away; similarly, by that time you will be surrounded by and breathing in oxygen atoms that started out from equally distant lands.

For an office worker in London that means he's surrounded by and inhaling oxygen atoms that had been in Paris a few days before, breathed out by men smoking Gauloises, or emerging from temporary storage in microscopic cavities in upholstery surfaces or wall paint there. In a year's time the oxygen atoms will have traveled a distance greater than the earth's circumference, and will be coming and going from every part of the globe. This is where things get interesting. Oxygen atoms last a very long time. Some of them are used up in all the paper attacking and other activities, such as being inhaled, but a good proportion stay available. About one-sixth of the oxygen atoms you breath in are released in the next breath, quite untouched by anything that went on inside. The same thing happens with everyone else who is breathing – your spouse, your boss, your friends, enemies and others – and because such terrific numbers of molecules are involved, because the stuff is so light and easily dispersed, it's statistically near certain that within a year's time you'll be breathing some of the oxygen, some of the exact same molecules, that they sucked in a year before.

We can extend the same reasoning to the past. A small sample of the oxygen molecules from *any* breath that anybody took within the past few thousand years is near certain to be in the next breath you take: oxygen molecules that were in the last gasp Lavoisier took; oxygen molecules that were fluttering around and then breathed in by John F. Kennedy when he was inaugurated, or by the Apostles listening to the Sermon on the Mount, or by unknown eighth century peasants hauling wood. It's worth pausing for a moment, then taking a deep breath and reflecting that samples of them all are in the air you have just drawn in.

What happens when the oxygen-assaulted book is downed will vary in different homes. In some, eyeglasses will be adjusted and the book will be replaced by a briefcase of office work lugged up. In others there will be sensible discussion of housework, and then lights out. But in yet other homes a distant memory of Monsieur de Calonne will be recalled, and, a reported average of 2.5 times a week for US married couples, activities will begin which set the bedsprings to shake, to jounce, to rebound and, as happens so often, so embarrassingly – and at just the worst moments – to enter into audible resonant frequencies and squeak.

Why do bedsprings do this to us? They are not being perverse, malicious, censorious, puritanical; they are just illustrating in concentrated form something that goes on whenever we apply weight to a solid surface. Take a step across your bedroom floor on the way to bed after demurely turning out the top light in the hope that this might be one of those 2.5 times and, unless there have been some shady deals in the zoning department, you are unlikely to fall through the floor. This is because as you apply your weight to your slippered foot, the floor, apparently passive and immobile beneath you, pushes back up with exactly as much force as you're applying down. This is miraculous. Touch your right toe to the floor in a mock-pirouette as you gingerly step from the bathroom towards the bed where your beloved awaits, exerting only a few ounces of pressure, and the floor will register that and push up with only a few ounces too; shift into a running scamper so that your full body weight is on that foot though and the floor, right in the same spot, will immediately rearrange itself inside and push up with the equivalent of your full body weight.

Nothing escapes its measure: the swiped scoop of mashed potato, that furtively devoured third helping of cake – the floor detects the weight of it all, and pushes up accordingly. It's important that it does so, for if it missed something you had eaten and pushed up too softly you would gradually start sinking through the floorboards, your slippers puckering out the ceiling below. If the floor over-estimated what you had eaten, then it would push up with more force than your body weight could overcome, and you would be propelled up like a trampolinist.

The floor knows all this because when you step on it you tell it how much you weigh. The more you've eaten, the more you crush the molecules in the floor beneath your eager slippers. This is the key. Crushed floor molecules push back. Each individual molecule in the segment of floor directly beneath your foot rebounds up to try to counterbalance the crushing. They can't push back very hard, each molecule being limited to a very small fraction of an ounce of rebound oomph because of its diminutive size, but as there are very many molecules in the floor the net result can be considerable. The more crush, the more they push back, so equalizing your weight – exactly. In a floor made of iron or concrete the push-back of the crushed floor molecules would be near instantaneous, which is why such surfaces feel hard underfoot. In wooden floors the crushed molecules respond with only a little more delay, and that's why such floors have a slight give but otherwise feel hard and solid when you jauntily skip across them.

Only for a metal bedspring it is different, for there the whole structure has been arranged in devilish spirals so that before the molecules get a chance to push back, before they get a chance to show how accurately and evenly they can assess the weight tumbling in on top of them, the whole structure of the metal will have deformed down. Only at the lowest point of that bending will the molecules finally get their big opportunity to wallop the weight overhead back up.

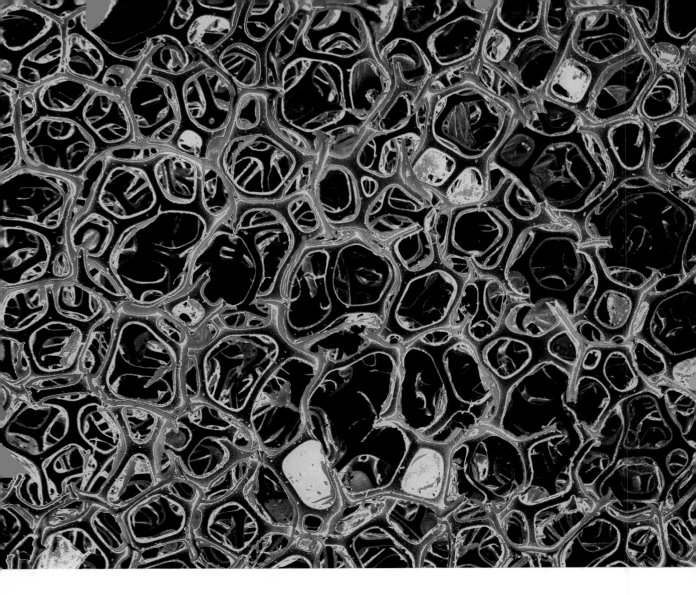

Which is a mistake. Bedspring molecules pushing from way down at the limit of deformation will not stop when they've reached the height they started off at. Rather they will have so much momentum, so much rocketing up velocity, that they will continue going, carrying whatever is on top with them. This is why springs bounce: they're doing just what an ordinary floor does, but more so. Now if this were all there was to it there should be little problem in engaging in rumbustious activity on the bed without turning the springs into an actively squeaking loudspeaker system. One would just have to be careful to minimize one's own oscillations, and the bedspring's molecules would minimize along with

Two solutions to the night-time resilience problem. Right: Tunneling down the spiral of a normal bedspring. Opposite: Interlinked cells in the plastic foam used in pillows and mattresses. The spring absorbs stress by deforming through its whole length; the plastic foam absorbs stress by deforming the delicate strutwork in the affected area alone

you. Alas, there is something more. A bouncing spring, however delicately moving, is a vulnerable object. When it compresses it stores all the energy of prior forces inside. Let an additional force strike it exactly in time with one of its bounces, and the spring will accumulate more force, and accordingly bounce up further on the next bound. A very little input will go a very long way. It's the same effect we get when by tapping a moderately swaying playground swing at exactly the right time it begins to go faster. Bedsprings cushion us when lying still, allowing more give than a hard floor would, but precisely because of that they have this characteristic of picking out and amplifying any additional force that comes down afterwards, if it's applied at the right frequency. And the modern bedspring has many such sensitive frequencies. Car springs, jetliner engine fastenings and other structures have the same propensity to fall into resonant frequencies, and only careful attention to the fundamental laws of oscillatory motion – attention more costly than bedspring manufacturers can afford – allows them to be overcome.

The man who first described these laws of spring action was Robert Hooke, a clergyman's son of multiple talents, who along with being a founder of modern science is responsible for the design of London, as he led the blueprint team after the City was burned in the Great Fire of 1666. Unfortunately these talents were alloyed with certain other traits. The kindest character analysis even a favorable biographer has been able to make is that perhaps if Boyle 'had not suffered from his ugliness and chronic ill health, he would not have been the difficult, suspicious, and cantankerous person he was . . .' Still, he became rich, and that, in English society of the time, allowed him to become an extraordinary philanderer. Servant girls, servant girls' female friends, his wife's female friends, his wife's friends' servant girls – all apparently had to be removed from his presence at various times because of his repeated and successfully forced advances. That he, rather than one of his more restrained colleagues, was the first to focus upon and analyze oscillatory spring motion, is perhaps open to biographical interpretation.

But, in time, the bedroom stills. Some condensed steam, some flung-loose cleansing cream, and a little heat in the now quiet springs is all that remains of the final evening labors. The time has come to sleep, or try to. Pillows get pressed, pummeled, adjusted and readjusted. Sheets are pulled, and final positions taken. There are gruntings, little fussing noises, some final reflex tugging and then, quite without noticing, silence. Until, that is, the torture begins.

Just as we are falling off to sleep there is a chance, an unfortunately good chance, that a certain repetitive and intrusive noise will be heard. Eyes closed, senses dim, the nature of the difficulty is at first hard to distinguish. There is something in it of a ping, a hint perhaps of a plop, and at that moment, unless one tries very hard to resist further analysis, the awful truth will soon become clear: it is a splat, a splat such as only one object in the house can make, one torturous device that the architect should be strangled for having dared to impose on you: the splat of a dripping bathroom tap.

Weaklings try to ignore it, but that is a mistake. Weaklings jam pillows over their head, screw their eyes tight, and hope that if they're very very good the awful noise will go away, or at least that their spouses will do something about it before they have to. But dripping taps do not go away, and spouses are rarely so foolish as to show signs of life at this moment. Dripping taps have no pity, extend no mercy, but just grow louder and louder, reverberating inside the attempted sleeper's head until he gives up, jams feet into slippers, and, cruelly woken, storms into the bathroom to deal with the culprit. If he is lucky the encounter with the tap will be a brief one, no bicep-straining battle with rusted metal, but a simple flick, a twist, and so a quick return. But if he is not lucky, there will be war.

The water droplet can work its way through even a tightly closed tap because it doesn't simply fall out in one glob, but rather dangles, stretches, pulls, tugs and only then snaps loose from the tap. This makes turning off a tap difficult. It is also why drops descend not in regular intervals, but in uneven gaps of teasing agony. Even once they come out the drops continue their contortions. The first thing they do when free from the tap is stretch out into something like an overextended rubber band shape. But then, since water has a certain amount of internal cohesion, it snaps back to the form of a wadded up rubber band, before rebounding out again for another stretch. It's like a high diver doing a series of tucks and stretches as he's in free fall. Only after a few cycles of stretch and snapping does the drop settle down to its ultimate appearance, which, due to all this unaerodynamic buffeting, is not the drop-shaped contour we might imagine but something resembling the profile of a hamburger bun. (With attentive peering you can just about see the drop's transformations.) Such internal deformation happens to the rubber in a tennis ball too, making the object of a McEnroe serve emerge flat as it leaves the racket, then rebound into a long furry sausage before pulling back into the original flat shape, then repeating this cycle all the way across to the opponents' receiving square, the ball's speed increasing and decreasing as it wobbles along. (This happens in every serve at Wimbledon, but the TV cameras, even slowed for instant replay, are too ungainly to detect it.)

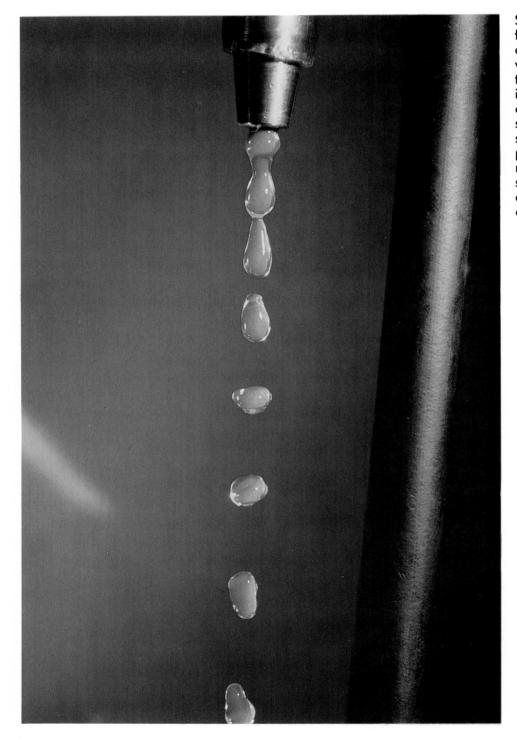

Stroboscopic freezing of a dripping tap; the water drop goes through incredible contortions of stretching and spreading, produced by air resistance as it speeds to make a rousing impact on the sink

When the water drop hits the sink it makes two separate sounds. To us the impact seems a brief, almost jerkily fast event, lasting as it does only $\frac{1}{100}$ of a second. But to the water droplet, whose inner molecules are used to bumping and interacting on a scale of mere billionths of a second, that impact movement is equivalent to many hours of slow shattering. There's time for writhing, bouncing, wiggling and even a quick shimmy, before the inevitable happens and the droplet explodes. If the explosion sent water fragments out at 180 miles per hour, all the energy of the falling would be used up in those mini-shrapnel displays, and there would be none left over to vibrate the sink and make a noise. But alas, falling water droplets in your sink have the bad grace to pop apart at a mere 140 miles per hour, and at that speed there's plenty of extra impact energy to send the sink twanging. If the basin were sculpted to resemble the inside of a cello the resultant sound would be perhaps as delicate as a base chord from a master. But as the sink your unimaginitive architect specified is probably a hackneyed thing merely sink-shaped in design – the acoustic equivalent of a deformed banjo – then all that the shrapneling droplet produces for the first part of its note is a discordant *ping* when it strikes.

The 140 miles per hour leftover fragments are what produce the second half of the sound. These fragments from the dismembering droplet create shock waves in the air in front of them as they float sideways away from the point of impact. Such shock waves on a larger scale are what thunder comes from; here in the sink the sound they produce is equally discordant, only softer. Combine the two effects and the full *ping-thunk* of a dripping tap is there.

This is what the homeowner has emerged from his bed to stop. He lunges at the tap and begins to squeeze, squeezing in wild, primitive jerks, ignoring the shearing of flesh from palm and squeezing till the last drop inside is severed, till there's not even the tiniest exit for others to seep out of, till he's sure there will be no more *ping*, no more *thunk*; he squeezes till the torture is over, and the tranquillity of his home is assured. Only then can he return to bed, where all is warm, and cosy, and quiet. Only then is the day, for our household, finally done.

Picture Acknowledgements

All photographs are supplied by The Science Photo Library, as follows:

Aga Infrared Systems/SPL 10; Professor Beidler/SPL 90; Dr Tony Brain/SPL 23, 43, 50, 86, 87, 92, 98, 110, 144, 146 (top and middle), 160, 211; Dr Jeremy Burgess/SPL 127 (bottom), 130, 146 (bottom), 156; Dr R.P. Clark/SPL 50, 111, 134, 163, 190, 194, 195; Martin Dohrn/SPL 22 (right), 86, 87; Dr H. Edgerton/SPL frontispiece, 222; Catherine Ellis/SPL 27 (top), 63; Vaughan Fleming/SPL 27 (bottom); Gennaro/SPL 51; M. Goff/SPL 50, 111, 134, 163, 190, 194, 195; Gorham/SPL 167; Grillone/SPL 51; G. Hadjo/CNRI/SPL 47; A. Hart-Davis/SPL 219; Elsa Hemming/SPL 98; Jan Hinsch/SPL 38, 142; Hutchings/SPL 167; Manfred Kage/SPL 59; Dr J. Lore/SPL 79; David Malin/SPL 107; Mike McNamee/SPL 14; Nelson Medina/SPL 174; Howard Metcalf/SPL 30, 31; Sidney Moulds/SPL 179; NASA/SPL 168; National Institute of Health/SPL 139 (top); David Parker/SPL 86, 87, 110; David Scharf/SPL 127 (top); Dr Gary Settles/SPL 74, 99, 138, 139 (bottom), 187, 199; SPL 15, 22 (left), 40, 41, 48, 58, 65, 68, 82, 97, 102, 116, 121, 124, 126, 150, 151, 154, 155, 204, 218; Don Thomson/SPL 54; Geoff Williams/SPL 30, 31.